THE
CLAUSEWITZ
DELUSION

THE CLAUSEWITZ DELUSION

How the American Army Screwed Up the Wars in Iraq and Afghanistan (A Way Forward)

STEPHEN L. MELTON

To Neil and Joseph Melton, my sons
who served in Operation Iraqi Freedom

First published in 2009 by Zenith Press, an imprint of MBI Publishing Company, 400 First Avenue North, Suite 300, Minneapolis, MN 55401 USA

Zenith Press titles are also available at discounts in bulk quantity for industrial or sales-promotional use. For details write to Special Sales Manager at MBI Publishing Company, 400 First Avenue North, Suite 300, Minneapolis, MN 55401 USA.

To find out more about our books, join us online at www.zenithpress.com.

Library of Congress Cataloging-in-Publication Data

Melton, Stephen L., 1952-
  The Clausewitz delusion : how the American army screwed up the wars in Iraq and Afghanistan (a way forward) / Stephen L. Melton.
    p. cm.
  Includes bibliographical references.
  ISBN 978-0-7603-3713-4 (hardcover)
  1. United States--Military policy. 2. Iraq War, 2003- 3. Afghan War, 2001- I. Title.

UA23.M463 2009
355'.033573--dc22

2009019051

Designer: Helena Shimizu
Cover Design: Simon Larkin
Design Manager: Brenda C. Canales

On the cover: During the early morning hours, Iraqi Police, supported by a U.S. Army M1A1 Abrams main battle tank, investigate an explosion-charred site caused by a vehicle-borne explosive device outside Gate Three of the Green Zone in Baghdad. *Staff Sergeant Ashley Brokop, USAF*

On the back cover: A U.S. Army M1A2 Abrams main battle tank of the 1st Battalion, 3rd Armored Combat Regiment, conducts a cordon-and-search operation in Biaj, Ninawa Province, Iraq, during Operation Iraqi Freedom. *Staff Sergeant Aaron D. Allmon II*

Printed in the United States of America

Contents

Author's Note

Throughout this book I constantly refer to the tactical, operational, and strategic levels of war, concepts that are confused in doctrine by circular definitions and rhetorical doublespeak—for instance, the ubiquitous reference to a "strategic corporal," a low-ranking soldier whose independent actions, for better or worse, change the fortunes of war. In this book I use the following set of definitions regarding the levels of war.

Tactical: All U.S. Army units—division headquarters, brigades, battalions, and lower-echelon units that rotate in and out of a theater of war. The army, in a sense, mass-produces these units and deliberately structures and trains them to perform specific, reproducible missions. A unit that can be replaced by another like unit is always a tactical unit. Soldiers who serve in these units make tactical decisions based on the guidance provided them by higher-level commanders. (All corporals are tactical. Corporals who exceed or disobey their orders are disciplined based on operational and strategic policy or statute.)

Operational: Three- or four-star general officer headquarters generally deployed overseas. The function at this level is to wage the campaigns necessary to win the wars ordered by the president and Congress. These headquarters are constituted for the duration of a war or even longer. Examples would be Multi-National Force-Iraq, Central Command, the Southwest Pacific Command in World War II, and the Army of the Potomac in the Civil War.

Strategic: The decisions made in Washington, D.C., by the president and his appointees, the Congress, or the Joint Chiefs of Staff. Strategic decisions include whether to go to war or make other military commitments; the manning, structure, budgets, and policies of the military departments; and the training, doctrine, and equipment provided to the fighting forces.

Grand strategy: I occasionally use this term to refer to enduring national strategies that transcend the strategic decisions made by any particular administration. For instance, "westward expansion" was an American grand strategy that lasted over two hundred years, until the frontier had been settled to the Pacific. "Containment" during the Cold War lasted over forty years, until the eventual fall of the Soviet Union.

I have also elected to use the term "military" to include all the American uniformed services: the U.S. Army, U.S. Navy, U.S. Marine Corps, and U.S. Air Force. This is the modern usage. I use "army" or "soldiers" when referring specifically to the United States Army or its members.

Preface

It has been my great privilege to serve in the United States Army for the past thirty-one years, first as an enlisted soldier, then as a commissioned officer, and finally as a civilian instructor at the army's Command and General Staff College. I have spent my entire adult life in and around the army because I love the army, I admire and appreciate soldiers and their families, and I believe that the army is among the great organizations in human history. The sacrifice and commitment of generations of American soldiers have brought freedom to countless millions, at home and worldwide, and have rescued countless others from unspeakable cruelty and oppression. A rare and noble combination of altruism, bravery, dedication, and competence sets American soldiers apart from their civilian counterparts and most other armies in the world. It is small wonder, then, that I have always preferred the company of soldiers.

But the army is not universally excellent. Nor is it always unerring. As an institution of human beings, it is flawed, sometimes seriously flawed.

The foundation of the army—the combat units that rotate in and out of the war zones in Iraq, Afghanistan, and other lesser-known commitments—is its best part. Though some combat units are better than others, the platoons, companies, and battalions that form the "point end" of the army are outstanding, comprising devoted young men and women working desperately to achieve their important and often dangerous missions. Everything that goes on at this level is profoundly honest and immediate. With no room for error, there is no tolerance for "bullshit." When you have to literally bury your mistakes, usually in flag-draped caskets and with full military honors, and live with your failures, everyone does everything humanly possible to get the mission done right. Brigade and division headquarters, though often a step or two removed from the combat or the actual work, still tend to be outstanding organizations. The heavy responsibilities of command and the close proximity to the troops in the field tend to make commanders and staffs at these levels fully engaged in their necessary functions. Though often chided by the soldiers of subordinate units as being "rear-echelon m—f—rs" who "don't know the deal," the reality is that the tactical and operational headquarters work long, hard hours trying to ensure the success and survival of the lower-ranking soldiers over whom they have command.

The top of the army, specifically the Department of the Army staff and the Training and Doctrine Command, is a vast bureaucracy sometimes referred to as the "Institutional Army." Its mission is not to rotate from stateside bases to missions overseas but rather to train, equip, man, and sustain the tactical units and operational headquarters that do. The Institutional Army is the strategic platform that enables the fighting units to succeed when ordered to war by the president and Congress.

This book questions whether the top part of the army did as well as it should have in conceptualizing and preparing for the wars that have consumed the nation for the past eight years. The book argues that the army entered the wars in Afghanistan and Iraq lacking a reasonable understanding of how they would or could be won, as events in both of these countries have proven. Perhaps worse, the army has embarked on a costly strategic military offensive without due consideration of the nation's chances of success and the very dangerous second- and third-order effects of such a strategy, which are also now becoming apparent. Most will find it astonishing that the best-equipped, best-trained, and most competent and professional army in history could have such difficulty understanding the very nature of war and lack a useful framework for evaluating potential grand strategies. But there it is; events speak for themselves.

In the army, perhaps more than elsewhere, the grind of immediate business at hand often takes away the time needed for reflective thought and future planning. The army has always been a very busy place, even in peacetime. The young men and women who become the officer leadership of the army work long and hard to master their duties and solve the myriad practical problems that consume so much of their attention. Officers get promoted, change jobs, or move to a new assignment at a new installation every year or two, meaning that these adventurous souls are constantly climbing a steep learning curve just to do their jobs day in and day out. By the time an officer finally learns his current job, he's already on orders to go to the next. Throw in a spouse and a few children—most soldiers want families—and every waking hour is consumed by the demands of work and family. There is no time for thought that doesn't have immediate payoffs.

Unfortunately, this pattern pervades all levels of the army, from the newest lieutenant to the most senior four-star general. Work all too often drives out thought. While in uniform I never had time to seriously research and reflect on the important questions confronting the army and the nation's military policy. Only my retirement from active duty, appointment to the faculty at the Command and General Staff College, and my children leaving for college have allowed me the time to develop the ideas and supporting evidence contained in this book.

Epiphanies such as mine, however, should not have to occur within the solitary realm of the rogue faculty member. In the vastness of the Institutional Army, there should be a place where critical strategic problems are thought through with due rigor. There should be a place where the professional body of knowledge of the army is maintained and updated, so that authoritative solutions to and recommendations regarding military problems may be provided to the Department of the Army staff, combatant commanders, and civilian leaders. Much as a faculty member in a college physics or chemistry department, for example, is responsible for understanding the history and current state of the department's body of knowledge, doing research to further the knowledge, and communicating the knowledge to students and practitioners outside of academia, so too must there be a single institution in the army directed and staffed to perform the same function regarding land warfare, especially at the operational and strategic levels. Some institution that has had time to think its way through all the data must be able to tell all the "busy" people in the rest of the army what has proved to be "true," what is "unproved," and what is known to be "false."

Unfortunately, the army has so many different agencies claiming some part of this mission that none of them speaks with authority or statutory responsibility. For instance, here at Fort Leavenworth we have the Command and General Staff College, the School for Advanced Military Studies, the Combat Studies Institute, the Combined Arms Doctrine Directorate, the Center for Army Lessons Learned, and other agencies with other "stovepiped" pieces of the puzzle. At Carlisle Barracks, Pennsylvania, there resides the U.S. Army War College, the Strategic Studies Institute, and the new Peacekeeping and Stability Operations Institute. The U.S. Navy, U.S. Air Force, and U.S. Marine Corps maintain similar institutions based on their services' requirements. Complicating matters, the chairman of the Joint Chiefs of Staff runs the National Defense University in Washington, D.C., and the Joint Forces Staff College in Norfolk, Virginia. All of these institutions, plus many others not mentioned above, provide benefit to the military and the nation, especially in their educational capacity. But the mixture is so confusing and overlapping that even the truly "busy" senior generals responsible for commanding the Institutional Army do not know where to go for authoritative advice regarding critical war-fighting knowledge. Lacking is that single academic agency responsible for seeing, conceptualizing, verifying, recording, and disseminating a complete and coherent picture of how wars are won at the operational and strategic level. This deficiency led to many of the army's initial setbacks in the global war on terror, and it risks even greater setbacks as the war proceeds. Correcting this deficiency should provide the army and the nation with a clearer path forward.

An army institution so constituted may reach conclusions other than the ones I present in this book. So be it. My goal in this book is merely to spark a debate about the true nature of warfare based on empirical evidence (generally historical) and scientific methods, insofar as possible.

Introduction

When I was a boy growing up in early 1960s, Americans boasted that we had never lost a war and confidently predicted that we never would. Our history of military-enabled expansion seemed a birthright—a proud tradition that my friends and I would inherit and fulfill when we became men. Successful American wars would continue to make the world a better place, not just for us but for all mankind.

Needless to say, things have not gone as we imagined they would. The backdrop of the Cold War, a rare interlude of strategic defense to consolidate our World War II gains, forced America into a period of small, frustrating conflicts in which the hoped-for fruits of victory rarely seemed worth the effort.

The demise of the Soviet Union in the 1990s after nearly five decades of the Cold War enabled the emergence of a "neoconservative" constituency for renewed American expansion, using military force as needed. The neoconservative vision that found voice in the presidency of George W. Bush was popularized by Thomas P. M. Barnett in his highly influential and best-selling work *The Pentagon's New Map* (2005). In this book Barnett called for the "functioning core" of modern nations, led by the United States, to expand their governance and economic principles to the backward, lawless, and misgoverned regions he characterized as being the unconnected "gap." The neocon argument is that bringing Western-style governance to these gap areas is necessary not only for their own improvement but indeed for the defense of the functioning core, the civilized world. Such thinking has led us into the current wars in Iraq and Afghanistan.

Despite America's overwhelming military firepower and unprecedented global reach, its excessive overmatch in air and naval forces, and its more than ample political and budgetary support, we are now bogged down in both of these countries, unable to conclude wars we began by our own choice over a half decade ago. The American citizenry, once very supportive of these wars, has become increasingly frustrated and disillusioned by the efforts. As victory remains elusive, defeatism grows.

Do we still understand how to win offensive wars? Has the U.S. military emerged from two generations of Cold War with a sufficient understanding of the nature of offensive warfare at the strategic, operational, and tactical levels, to enable the neoconservative urge to renew America's historical expansion? Sadly, the answer is no. A casual observer would have assumed that the U.S. Army, now retasked by the executive branch of government for offensive wars, would have studied its prior successes and resumed its earlier practices, picking up from where we left off in the 1940s. But we didn't.

The U.S. Army is a huge organization, now subordinate to an even larger Department of Defense—a Cold War moniker if there ever was one. Both organizations were paralyzed in the 1990s and into the twenty-first century by an overwhelming institutional inertia of rest, trapped by Cold War defensive paradigms unsuited to unforeseen offensive missions in Afghanistan and Iraq.

From a personality perspective, army officers tend to be introspective when they collect the data they need for decision making. Rather than look at the outside world and draw conclusions based on analyses of external reality, our dominant psychological preference is to view the world in terms of our own personal experiences and passionately defend our past actions.[1] This psychological phenomenon explains why the army is so tradition bound, backward looking, and slow to adjust to changing realities: generals are always accused of being eager to fight the last war rather than preparing to fight the next war. When America was constantly fighting offensive wars—that is, from the initial colonization through World War II—the American military transmitted knowledge of offensive warfare from generation to generation, officers usually perfecting in middle age the skills they initially learned in their twenties.

For nearly three and a half centuries, the U.S. Army's traditions appropriately prepared upcoming generations for the offensive wars that they would fight as America expanded from two small colonial settlements into the world's dominant power. But America's victory in World War II inescapably brought on a prolonged, multigenerational period of defensive warfare heretofore unknown to American soldiers. Successful and brilliant army officers of the offensive era—for example, Gen. Douglas MacArthur and Gen. William Westmoreland—struggled and ultimately failed to reconcile their offensive war upbringings with the imperatives of the new national strategy of containment, which sought to defend only what had already been won.

As the Cold War progressed, however, new generations of officers steeped in defensive warfare gradually replaced the World War II generation. In the late 1970s the officer corps emotionally abandoned our failed and overreaching war to bring democracy to Vietnam and in turn embraced the prospective defensive war in Europe—against the looming behemoth of the Soviet Union's Red Army—making it their generation's commitment to preserve freedom only where it already existed. Successive army doctrines dating from the post-Vietnam period—first firepower dominance, then Active Defense, and finally AirLand Battle—affirmed the American military's belated acceptance of a defensive national policy enunciated decades earlier by George Kennan, Harry Truman, and even Dwight Eisenhower.

Over time, the conversion of the army rank and file to a defensive mentality would be complete. Numerous on-the-ground and schoolhouse rehearsals of the defensive plans for West Germany and South Korea, plus yearly unit training rotations to the National Training Center at Fort Irwin, California, where the capstone exercise was always the unit's defense against an attacking, Soviet-styled regiment, had convinced anyone in uniform whose careers had survived these peacetime challenges that being able to defeat an attacking enemy was really all that mattered. The post-Vietnam Cold War army had irrevocably wedded itself to defensive warfare about thirty years—one generation—after the political decision had been made in Washington to adopt a defensive grand strategy.

Hardly a decade more had passed, however, when the Berlin Wall fell in 1989, unsettling the army's newfound defensive groove. Senior officers were slow to adjust. In 2004, fifteen years later, they would still begin their Command and General Staff College (CGSC) presentations with the lame observation that the Cold War was over, even though the officers they were addressing were too young to have any personal recollection of that struggle against Communist expansion. The thinking of the institutional leadership was at least ten years behind the times.

Even in the early 1990s, when the neoconservatives began pressing for a new offensive military strategy to exploit our Cold War victory, the collective response of the military was to pooh-pooh nation-building as something *we* don't do.

Oh, really? It was as if we could not even bother to open up the history books to understand that throughout the recorded history of our own army we had successfully conducted offensive warfare *and* nation-building for the first 350 years of our existence as an identifiable people.

By 9/11, after a generation of concentrated training on defensive battle, offensive warfare simply no longer existed in the collective memory of the army as an institution. No one had experienced it, so, as introspective reflectors, we corporately denied its existence. Now, in our sixth year in Iraq, we still deny its existence; we imagine we are defending a host nation.

Perhaps worse still was the army's post-Vietnam descent into postmodernism, a belief that the old formulas had become passé and that new, more imaginative, innovative, and intellectual approaches to military problems were required. Our first ever defeat, in Vietnam, caused the American military to devalue the lessons of our military tradition, despite its manifest long-term success, and search for new, outside-the-box solutions. Struggling to understand their defeat, the generation of officers who entered the army during Vietnam, the same officers who would eventually be the generals in Afghanistan and Iraq, collectively became enamored of the thoughts of Carl von Clausewitz, a German military philosopher from the early nineteenth century.

In the late 1970s Clausewitz's recondite *On War* became the army's new intellectual touchstone, viewed as revealed wisdom from a more sophisticated European military tradition. Having lost faith in ourselves in the 1970s, the army of the 1980s replaced William Tecumseh Sherman with Carl von Clausewitz as the primary author of the army's philosophical foundation. In the blink of an eye, "center of gravity analysis" and "decisive battle" replaced "attrition" as the basis of the army's understanding of war.

Reinventing ourselves as nineteenth-century Prussians was perhaps the worst way to resolve the army's post-Vietnam crisis. In retrospect, this empathic attachment to a distant mind all too often prevented the army from making the necessary and more relevant examination of our own American successes and failures.

But the most disturbing quality of late-twentieth-century army doctrine is its intensely ahistorical and unempirical understanding of warfare, its implicit rejection of the concept that there are any patterns in warfare worth studying. The post-Vietnam army imagines itself in an ever-continuing revolution in military affairs that makes the study of past wars irrelevant to the decisions we must make today. Indeed, Clausewitz, too, was a historicist, believing that each battle was so unique that trying to analytically discern patterns among them was useless.

Rather than take the rationalist, engineering approach to war that characterized the U.S. Army's experience, Clausewitz explained success in battle in terms of intangibles: the genius of the leader, the role of chance, the will of the army, and its mood for battle. Reflecting these rediscovered romantic-era notions, U.S. Army doctrine in the 1980s and 1990s increasingly viewed all situations as being inherently different, and it challenged commanders to come up with unique solutions for each problem they faced; after all, in a Clausewitzian universe, each problem is unique. The problem with this approach is that it relies on finding *genius* in combat leaders at the expense of giving merely good commanders a way to routinely achieve 80 percent solutions to the problems they face. And "genius" is rare.

If we deny that patterns exist in warfare, we won't look for them. Then art transcends science as the cornerstone of military understanding. Indeed, in today's army, commander-centric operational art has replaced military science as the doctrinal means of solving military problems.[2] The pragmatism and engineering approaches that characterized the army of the Progressive Era and World War II have been abandoned. Instead, the army now embraces the intangibles of art and intuition, all the while hoping that military leaders can somehow derive their own formulas for success.

As a consequence, the nitty-gritty of our nation's military experience is missing from our doctrine and consequently our operational and strategic thinking. It wasn't that our rich historical experience has been summarized into bland institutional prose but rather that the lessons of the past have been ignored altogether. How many of the enemy will we have to kill, how many civilian casualties and how much economic destruction will we have to inflict, how many soldiers will we need, how many will we lose, how long will the war last, how much will it cost, what can we reasonably hope to achieve? These critical strategic questions in warfare are no longer analyzed in our doctrine or professional education. The histories are rich in these details, but our published doctrine is generally devoid of analysis or meaningful conclusions regarding these critical strategic issues. Consequently, we in the military find it difficult to intellectually frame wars: their costs, their benefits, their very nature.

The U.S. Army's disregard of the historical record frustrates efforts to understand the nature of current or contemplated wars. An exchange between the author and Dr. Richard Stewart, the chief historian at the U.S. Army Center of Military History, at the September 2008 U.S. Military History Symposium at Fort Leavenworth, Kansas, illuminates the lost opportunities to learn from historical experience:

> The Author: I was just reading the new Field Manual (FM) 3-07, "Stability Operations," which is excellent, and in Chapter 5 they talk about "transitional military authority." If you look at the previous version of FM 3-07 it was not there, no analogous chapter. If you read FM 3-07, Chapter 5, much of it is a lift out of FM 27-5, "Military Government," that we published in 1940 before we were even in World War II. It was based on our progressive era experiences after the Spanish American War, and the occupation of the Rhineland after World War I. My question to you is, as a historian, a military historian, why does the Army forget its history? Why are we relearning what we once knew?
>
> Dr. Stewart: That is an excellent question and one that I have continually wrestled with as an officer and as a civilian and as a historian because it . . . maybe I am predisposed to history, to think historically, to ask the question occasionally that is asked even in the Pentagon, has this ever happened before? Maybe I ought to look into this. If I could get people to ask one question when they are beginning to plan for an operation, that would be the one because at least it would get them thinking about the possibility that there might be information out there that might be discoverable and might be useable. The transition part, in

particular, is interesting. Everyone talks about lack of Phase IV planning for Operation Iraqi Freedom, right? Oh, we did not do any Phase IV planning. I was in the 352d Civil Affairs Command at the time with planning for postoccupation operations in Iraq, and we had six phases in our particular plan. We had Phase IV, yes, which was immediate postemergency reconstruction activities, but we went on in our plan and continually thought about Phase V, how do we begin to pull out of this and turn things over to Iraqi officials, and then Phase VI, close things out completely, sort of lock and head out the door as a civil affairs military government type organization? That entire plan with all six phases which dealt with transition was not used at all by CENTCOM when they decided to basically throw the plan out and start all over again. Not only do we not use history from Spanish American War, World War I, Rhineland occupation, and by the way they did not look at that report from 1920 when it was produced until 1940. So there was 20 years where they had collective amnesia. Not only did they not look at the far past, to look for appropriate historical examples, they do not even look at the more recent past, and the more recent possibilities, their own historical thinking and plans that they are generating now because they cannot be bothered to think about that old stuff. I remember briefing General Potter once from Special Operations Command on the Haiti Operations and I wanted to give him some sense of the problems that the Marines ran into in Haiti in the 1920s and 1930s. I had just started on my presentation full of historical beans and he says, "Look at those years, 1919–1938, that was 50 years ago; sit down." Maybe that is more Dick Potter than anyone else, but the fact is, he was not willing to listen to old historical experience because to him it had no application to today's current situation. If we could go back to inculcating in our officer corps, not just in the Army but in other services as well, some of the historical mindedness that was so much a mission, so much a part of TRADOC in the 1980s and 1990s. You remember, Sir, a battle analysis and a tied [*sic*] year long process, a historical analysis as well as teaching by the Combat Studies Institute in an attempt to inculcate in every officer a sense that history is alive, that history is useful, that you need to learn it and know where to find it, and know it has continuing validity. Pretty important lessons. If we could go back to some of that, perhaps that would help. Again you got me on my soap box. Thank you.[3]

The army's corporate failure to adequately consider the historical record has resulted in dangerous flights of intellectual fancy in U.S. military doctrine and wartime planning. For example, after the Vietnam War, the army adopted the belief that body counts were a meaningless statistic in warfare, despite the fact that insofar as was possible all of our military forefathers had taken great pains to obtain and record this basic battle data.

This new formulation of a fundamental reality of warfare seems at odds with the broad and successful American military tradition that over the course of three centuries transformed a few small colonial villages into the most extensive empire in history. Offensive wars have almost always been won by attrition, at least in the American historical experience. Kill ratios based on body counts do matter, not only in the outcomes of battles but in the outcomes of wars. Inflicting casualties on the enemy he cannot sustain while lowering friendly casualties to the point that they are sustainable is a necessary condition for victory in modern war. To throw out centuries of valid data based on a single example, in this case a flawed reading of the dynamics of the Vietnam War, was a tremendous intellectual error.

It seems to me that doctrinal changes should reflect the incorporation of new factual data into the existing body of knowledge, not have as its intellectual basis a disregard for the previous facts. Any doctrinal rewrite should be able to cite its empirical basis and explain how the new doctrine results in a better synthesis of all the available information. Anything else is suspect.

In yet another emotional overreaction to Vietnam, U.S. Army counterinsurgency doctrine fell into disrepair, and the mindset developed that general-purpose forces would never perform counterinsurgency again; special forces could handle it. The general tenor of any discussion on counterinsurgency seemed to imply that the tactics used in Vietnam were somehow extreme, yet any fair reading of American history, or any survey of the international experience for that matter, suggests that "hard-war" policies are necessary when conducting counterinsurgency campaigns, especially by an alien force in a foreign country. Hard-war counterinsurgency, however, is among the uglier forms of attrition warfare. It fell out of vogue in the smaller, all-volunteer army, so its realities went unconsidered and unstudied for the three decades between the wars in Vietnam and Iraq. Only years of frustration in Iraq have caused the army to update and republish its counterinsurgency doctrine.

Similarly, the army collectively overreacted to the failure of our military advisors in Vietnam to put more spine into the Army of the Republic of Vietnam (ARVN), not to mention the perceived career-damaging consequences of advisor assignments, and failed to institutionally support a robust advisor capability after the war, again relegating the task of building foreign security forces to a small corps of special forces. We are now rediscovering that assisting foreign armed forces is a key U.S. Army competency in the aftermath of offensive warfare as we desperately, though inadequately, try to create from a standing start thousands of advisors. Was this not foreseeable? Who did the army think would reorganize, retrain, and supervise the security forces of the nations we defeated in our strategic offensives of the global war on terror?

A victim of a Cold War defensive mindset (not specifically the Vietnam experience), army doctrine concerning military occupation of foreign territory, martial law, and military governance had fallen into disuse and been declared obsolete. The schools needed to train occupation troops and the military units that would be needed to control and govern occupied territories were first downsized, then eliminated, and ultimately forgotten. The post-Vietnam Cold War army felt that we would never again conquer a hostile foreign country and build up a more agreeable nation in its place. Disregarding 150 years of American military success in governing conquered peoples, we kidded ourselves into believing that if there was nation-building to do, some other institution would do it, not the army.

During my army career, as I struggled to glean practical knowledge from our official publications, I sometimes imagined that there was a guarded vault somewhere, maybe at the Army War College, maybe in the Pentagon, where the real truths about warfare were recorded. I imagined a trusted army staffer entering a secret, dimly lit library, reverently pulling ancient leather-bound tomes off the shelves, carefully wiping the dust off the covers to reveal titles and subjects, slowly fingering through the pages of time-honored truths to reach the passages regarding the questions at issue, and then confidently bringing the wisdom of the ages to the decision makers at the seat of power.

But soon I realized that my hopes of recorded truth were mere fantasy. The army did not maintain a library of important facts that scribes meticulously updated as additional data came in. There were no data series, no moving averages, no historical analyses to give us confidence in our current decisions or projections into the future. Consequently, the army forgot what we once knew.

We, the post–Cold War army, were just making it up, writing plausible narratives regarding warfare only loosely related to some obscure set of facts, much like television scriptwriters create story lines loosely based on some recent news event. When it came to our critical doctrine—how do we win wars?—we were institutionally weak. The weakness remained unexposed only because in the relative peace of the late Cold War there was no one to prove us wrong.

The insufficiency of our doctrine explains the tremendous intellectual uncertainty that gripped the army in the run-up to Operation Iraqi Freedom. There were no chapters in any of our manuals entitled "How to Conduct a Military Occupation" or "How to Install a New Government in an Occupied Country," even though the U.S. Army has a long and successful history of conducting these types of operations. Suddenly in 2001, occupation and governance had become important issues, but the U.S. Army as an institution was unprepared to meaningfully advise the administration as it formulated its war policies.

There was a vague though by no means unanimous feeling in some military quarters that we weren't sending enough troops to occupy Iraq, but we had no written occupation doctrine, much less one based on data and analysis that would convince civilian leadership that our advice was anything more than professional opinion. The army's inability and reluctance to "tell truth to power" in 2002 reflected the weakness of the army's institutional understanding of the kind of warfare that the nation had just ordered it to win.

General Eric Shinseki's late February 2003 statement to Congress that several hundred thousand troops might be needed to occupy Iraq was not only a year too late—the war would start three weeks after General Shinseki's testimony— but was also out of step with then-prevailing army doctrine. Very late in the day, it was as if the army was admitting that it had lost sight of something that was critically important.

In Iraq the army's contemporary speculations on war failed their first reality check: we went to war with no occupation doctrine, no nation-building doctrine, no army organizations specifically designed and trained for occupation duty, no advisor corps to rebuild Iraqi security institutions, no plan for procuring the necessary legions of translators, no institutional understanding of Arabic culture, and no counterinsurgency doctrine. Not even realizing that what we didn't know was important, the U.S. Army blundered into Iraq and created chaos.

Our professional responsibility to the nation was to know our business, to maintain the professional body of knowledge and keep it continually pertinent, to provide authoritative advice based on irrefutable data, and to be ready, given reasonable prior notification, to assemble the right mix of forces necessary for victory. In all these respects we failed, mainly because we had forgotten hard-won lessons from our history.

This book is an attempt to rediscover a framework for understanding warfare that is based on the enduring truths recorded across the broad historical record. A new framework is needed because Iraq has exposed our existing intellectual architecture—the neo-Clausewitzian doctrine of Full-Spectrum Operations—as being incomplete, inappropriate, and deeply flawed. Iraq is not an anomaly, not an exceptional case that would challenge even the most brilliant military doctrine. Rather we have failed in Iraq because the army no longer understands how offensive wars are won. Indeed, our prevailing military paradigm, a direct descendent of Cold War thinking, does not even recognize that such a thing as offensive war exists.

Consequently, we no longer have the doctrine, force structure, or training programs necessary to execute offensive war. We no longer possess even the vocabulary necessary to discuss the concepts central to an understanding of offensive war. Were we on the strategic defensive, as we were during the Cold War, this book would be little more than an academic treatise. But now that

America is again engaged in a strategic military offensive, whether one agrees with that strategy or not, it is vital that we understand offensive war better than we do now, not just understand it better but realize what is required for victory. We *can* and *must* do better—much better.

There are a goodly number of books that discuss why we have failed to establish favorable indigenous governance in Iraq. Most concentrate on the administration's errors or the legacy of Saddam Hussein or the various failures of the Iraqis themselves. I argue that the fault is equally shared by a U.S. military so narrow in its vision of warfare that it can't bring itself to recognize the dynamics of offensive war and meaningfully organize for it. This book continues the discussion begun in Barnett's *The Pentagon's New Map*[4] but reaches different conclusions. I explain why the entity that must take responsibility for military governance after offensive wars must be the U.S. Army rather than some multinational, interagency SysAdmin force that Barnett or the State Department would have us create.

In the appendix I have included the main text of FM 27-5, the army field manual on military government published in 1940, which is the best single statement I've found on post-offensive war civil affairs. It led the army, institutionally, to the successful occupations of Germany and Japan. As with the rest of our offensive warfare doctrine, an understanding of what worked in the past is far more useful than advancing untested ideas with the hopes that they'll succeed.

Chapter 1

America's Neo-Clausewitzians

Carl Phillip Gottfried von Clausewitz (1780–1831) served in the Prussian army throughout the Napoleonic Wars and served as director of the Kriegsakademie (War College) in Berlin from 1818 to 1830. His magnum opus, *On War*, was published after his death by his wife; his untimely death from cholera prevented final editing.[1]

The book is remarkable for its impact on German military thinking. Generalfeldmarschall Helmuth Graf von Moltke, who was chief of staff of the Prussian army from 1857 to 1888, personally promoted *On War*, ranking it with the Bible and works by Homer in importance. Alfred Graf von Schlieffen, chief of the German general staff from 1891 to 1905 and the great architect of Germany's World War I strategy, personally wrote the laudatory introduction to the 1905 edition. In 1933 Generalfeldmarschall Werner von Blomberg, Hitler's defense minister and eventually the Wehrmacht commander, wrote, "Clausewitz's book *On War* remains for all time the basis for any rational development in the art of war."[2] Even today at the German General Staff College in Hamburg, the Clausewitz Society sponsors a weeklong symposium on the life and ideas of their namesake author and ceremonially presents each student officer with a finely crafted, leather-bound edition of *On War*.

Some of Clausewitz's key ideas remain strikingly relevant. Leaders should understand the nature of the war that they are about to embark on before they begin it. War is a political instrument, its prosecution always linked to its purpose. Fog, friction, and chance all play their roles in determining battle outcomes. Clausewitz's many dictums and rules for winning various types of engagements and battles are valid and remarkably similar to those found in the tactical manuals of contemporary armies around the world.

Although many of Clausewitz's observations may rise to the heights of eternal truth, he was still a mortal and temporal being and, to that extent, a captive of his time and place. Although he sometimes alluded to classical battles, his references were nearly always to the late dynastic wars, especially of Frederick the Great, and the Napoleonic Wars: wars of European kings and emperors. For Clausewitz, other forms or patterns of war simply did not exist or, perhaps, just did not merit inclusion in his 660-page study.

The Clausewitzian vision is monarchical, and the unity of purpose between the monarch and the army is absolute. Politics is seen in monarchical terms: questions of alliance, border provinces, international interests, and tribute. Decisively defeating the opponent's army in battle is the means to settle the geopolitical issue.

> Sometimes the *political and military objective is the same*—for example, the conquest of a province. In other cases the political object will not provide a suitable military objective. In that event, another military objective must be adopted that will serve the political purpose and symbolize it in the peace negotiations.[3]

The result of the war is never final; the defeated king will find means to redress his losses by forming new alliances and raising new armies, preparing for yet another round in a never-ending political and military contest among princes.[4] Politics and war are both viewed as natural components of the dialog among monarchs.

> We maintain, on the contrary, that war is simply a continuation of political intercourse, with the addition of other means. We deliberately use the phrase "with the addition of other means" because we also want to make clear that the war itself does not suspend political intercourse or change it into something entirely different. In essentials that intercourse continues, irrespective of the means it employs. The main lines along which military events progress, and to which they are restricted, are political lines that continue throughout the war into the subsequent peace. How could it be otherwise? Do political relations between peoples and their governments stop when diplomatic notes are no longer exchanged?[5]

Indeed, much of *On War* is a philosophical attempt to resolve the dialectical contradiction between the unlimited violence of battle and the very limited political objectives of monarchical war. The author constantly reminds his readers to subordinate emotional wartime responses to rational political calculations, as difficult as that might be in practice.

The populace, whose duty is to obey their royal masters and patriotically support their wars, is nearly absent from the Clausewitzian calculus, meriting only a few brief passages, despite the democratic tide then sweeping in from the Atlantic. French patriotism and universal conscription are credited with providing Napoleon a much larger army than had previously been imaginable, but to Clausewitz the French army was still an army working for an emperor.[6]

The brief five-page chapter entitled "The People in Arms" discusses the possibility of militias and irregular forces harassing an invading army as a possible form of defense, but it avoids drawing conclusions because, Clausewitz observed, "this sort of warfare is not as yet very common; those who have observed it for any length of time have not yet reported enough about it."[7] His only advice regarding what we now call counterinsurgency is, "Where the population is concentrated in villages, the most restless communities can be garrisoned, or even looted and burned down as punishment; but that scarcely can be done in, say, a Westphalian farming area."[8] There is no discussion of post-conflict governance or what current U.S. military doctrine terms "stability operations."

To Clausewitz, the people would dutifully support the nobles and monarchs to whom they were bound, recognizing that the affairs of state were not the affairs of the populace. Monarchs would play; people would obey.

Clausewitz was a Prussian and viewed his world through that geopolitical lens. Revealingly, Clausewitz concludes *On War* by war-gaming a hypothetical allied attack and defeat of Napoleon two decades before, seemingly to see whether German alliance could have succeeded in defeating France without Russian help. In his analysis, the French center of gravity is Paris and its defending imperial army. The main concern of the allied commanders in chief must be to seek the necessary major battle on the approaches to Paris and fight it with sufficient superiority in numbers as to promise decisive victory.[9] The fewest possible men should be diverted to other purposes. Prussia and Austria are seen as the key allied players. "Theirs is the genuine striking-power, theirs is the strongest blade. Each is a monarchy, experienced in war. Their interests are clearly defined; they are independent powers, preeminent above all the rest." The more liberal, hence inconsequential, German federation along the Rhine would have to serve subordinate to the Austro-Prussian nucleus.[10]

By destroying the French army and capturing Paris, "France can be brought to her knees and taught a lesson any time she chooses to resume that insolent behavior with which she has burdened Europe for a hundred and fifty years. Only on the far side of Paris, only on the Loire can she be made to accept the conditions which the peace of Europe calls for."[11] Clausewitz's hypothetical plan for conquering France animated German military strategy for more than a century.

Clausewitz's strong preference for decisive, war-ending battles of annihilation, as opposed to protracted wars of attrition, reflected the Prussian strategic predicament. No matter how finely honed its military strength, Prussia was nevertheless a small country surrounded by more populous empires capable of fielding larger armies. Long wars did not favor Prussia. Trading soldiers in indecisive battles favored Prussia's enemies. These were the background assumptions to Clausewitz's scholarship at the Kriegsakademie. Clausewitz demanded decisive battles and rapid campaigns, directly attacking the enemy center of gravity to destroy the enemy army and thereby decide the political question in one great climactic bloodletting. "Shrinking from this battle [that is, accepting protracted wars of attrition] carries its own punishment."[12]

Importantly, Clausewitz counsels that the attacker's investment in blood will be redeemed by the moral destruction, pursuit, and annihilation of the defeated defending army. From the Prussian monarch's perspective, arranging and winning the great battle, whatever its cost, was more important than his army's efficiency in battle. In attrition warfare, of course, the opposite is true: relative efficiency counts for everything.

In hindsight, Clausewitz's theories, whether original to him or merely reflective of the Prussian military consensus, proved disastrous for the German nation and Europe, failing to accurately predict the outcome of its three empirical tests: the Franco-Prussian War, World War I, and World War II. (It is indeed noteworthy that all three wars began with the attack on France that Clausewitz envisioned in the conclusion of *On War*.) The September 1870 Prussian victory at Sedan, a testament to superior German mobilization, equipment, and strategy, indeed demonstrated that Prussian armies could achieve complete military victory in the sought great battle. However, the capture of the French emperor Napoleon III by the Germans unleashed republican popular sentiment throughout France, giving birth to a new French government determined to continue to resist the German aggression. To Moltke's surprise, new French armies were raised and Paris and Metz defended against German occupation, if only for a few months. Only in January 1871, after the surrender of Paris, did the new French government agree to an armistice. But as the Germans, per the agreement, marched out of Paris in early March, radicals took over the Paris Commune, refusing to recognize the French government or its peace with Germany.

In the civil war that followed, French troops had to crush the Commune to restore order. In the aftermath, monarchical rule in France collapsed and the French initiated the Third Republic. Although the Prussians had succeeded in their limited war aims and annexed the Alsace-Lorraine, in a greater sense the Franco-Prussian War signaled to the world that the people and their representatives, not kings and emperors, would determine questions of war and peace in the new democratic age. This war also left a lasting French bitterness that would seek its revenge in a much bloodier Franco-German war two generations later.

The two world wars provide even greater evidence of the unsuitability of Clausewitzian theory in the modern age. Despite Germany's tactical and operational excellence, despite its offensive fighting spirit, despite its unity of purpose between sovereign and army, despite its spectacular initial victories in battle, both wars ultimately turned into wars of attrition that Germany eventually lost. The illusory decisive battle and allied capitulation never occurred. Each new German victory only steeled the democratic allies to even greater heights of collective resistance. Predictably, as the wars progressed, the allies gained strength while Germany weakened. The combined military, civilian, and economic costs of these lengthy wars of attrition brought Germany to its knees twice within thirty years.

To make matters even worse, the German army's harsh occupation of conquered lands, reflective of its undemocratic military doctrine and antirepublican politics, made the German army the most despised military organization of the twentieth century, unleashing popular resistance everywhere it went.[13] To this day, the Nazis and their armies deservedly remain the iconic evil in popular culture. Clausewitz had it wrong, and even to Germans the outcomes of the world wars demonstrated that Clausewitzian seed yielded a bitter harvest.

Introducing Clausewitz to the U.S. Army

It is indeed remarkable that three decades ago the U.S. Army, with its proud and successful heritage of democratic warfare, would scrap its prevailing doctrine in favor of the specious theories of a German army that we had soundly defeated twice. For 150 years, until the late 1970s, Clausewitz's writing was virtually unknown in the United States. English translations even during the world wars were poor and virtually unreadable. *On War*'s ponderous and obscure dialectical style, so characteristic of early-nineteenth-century German philosophy, was sure to put off any but the most academic American reader. The U.S. Army, steeped in American pragmatism and enjoying the success of its own military traditions, had neither the inclination nor the need to contemplate Clausewitz's deeper meanings.

The appearance of a better English translation of *On War* in 1976 (translators Michael Howard and Peter Paret, Princeton University Press), while America was struggling to understand its agonizing defeat in Vietnam, was the coincidence of events that against all odds propelled Clausewitz into the center of American military thought. The Michael Howard and Peter Paret translation of *On War* may have died an obscure death, like so many scholastic volumes gathering dust in so many libraries, had it been published at another time.

The new Clausewitz translation was not an overnight bestseller. The library at the U.S. Army Command and General Staff College (CGSC) at Fort Leavenworth, Kansas, bought just two copies, to set on the shelves alongside their two copies of the 1950 Infantry Journal Press translation and their four copies of the 1911 Graham translation.

The U.S. Army's infatuation with Clausewitz did not begin from within, but really spread to us from our brethren in the navy and air force. The Howard and Paret translation became a primary textbook at the Naval War College in the year it was published and was adopted by the Air War College in 1978. It took until 1981 for *On War* to become mandatory reading at the Army War College.[14] It was there that faculty member Col. Harry Summers melded Clausewitz's ideas with his experience in Vietnam and wrote his own book, *On Strategy: A Critical Analysis of the Vietnam War*, published in 1982.[15]

Clausewitz is the featured star of this acclaimed work, quoted dozens of times, almost on every page, much like a preacher would cite scripture. The analysis, in summary, is that the North Vietnamese Army, not the Viet Cong, was the enemy center of gravity against which the U.S. military should have focused its strength. South Vietnam could have survived if only we had concentrated our effort against North Vietnam by perhaps, Summers suggests, cordoning Indochina along the demilitarized zone.[16]

Whether Clausewitz himself would have agreed with Summers' solution is open to question. The very opening of the Introduction to *On Strategy* begins with a personal vignette in which Summers, then part of an American team negotiating peace with the North Vietnamese, discusses the war with his counterpart, Colonel Tu of North Vietnam.

> COLONEL SUMMERS: "You know you never defeated us
> in battle."
> COLONEL TU, after some pondering: "That may be true, but
> it is also irrelevant."[17]

How Summers proceeds from that introduction to an acclamation of Clausewitz, the high priest of the decisive offensive battle, is testament to the intellectual confusion that attended our losing the Vietnam War, America's and its army's first-ever defeat.

There can be no doubt, however, that Summers made Clausewitz, and especially center of gravity analysis, the new craze in American military thought. Suddenly, everyone in the military science community was reading *On War* and arguing about what it meant. Overnight, Clausewitz became mandatory reading for the faculty and students at Fort Leavenworth's CGSC and the Army War College. In 1984 Princeton University Press had to reprint the Howard/Paret translation, and the CGSC library upped its holdings to almost two hundred copies. In 2007 the Command and General Staff College bought over a thousand new copies of the 1989 edition and now has nearly as many copies of Clausewitz as it has students to read them.

Beginning in the early 1980s, Clausewitz and *On War* invaded U.S. Army capstone doctrinal publications, its written explanations defining the reality of war. There is not a hint of Clausewitz in the 1968 or 1973 versions of Field Manual 100-5, "Army Operations," or in earlier editions dating back to World

War II. The 1982 edition, the first of the AirLand Battle series, was the first to quote Clausewitz, though only once; however, *On War* did merit a featured mention in the brief bibliography. It is the 1986 edition of FM 100-5 that first gushes Clausewitzian theories, citing him four times. (Only Gen. George Patton earned more quotations, with five.) Here the U.S. Army learned for the first time in its then 211-year history that there is an "operational art" and that identifying the enemy center of gravity is its essence.[18] In appendix B, "Key Concepts of Operational Design," FM 100-5 paraphrases Clausewitz in its explanations of centers of gravity, lines of operation, and culminating points. The 1993 FM 100-5 added paragraphs on "decisive points" and "will" to its chapter on operational design. The 2001 version of Army Operations, now renumbered FM 3-0, included "center of gravity" in fourteen different discussions.

The latest FM 3-0, written with the army's failures in Iraq squarely in mind, features a tortuous discussion of center of gravity (now an acronym: COG), again the central idea in operational art and design. Here the army is told that the center of gravity may be physical or moral, military or nonmilitary, single or multiple, and that it may shift and adjust. Still, we are told that the COG is a vital analytical tool, that misunderstanding of the COG can adversely affect operations, and that the COG may be approached directly or indirectly. Decisive points are not COGs, but their loss weakens a COG, even though the COG is the source of the enemy's power and strength.[19] (Pity the poor staff officer forced to make sense of this gibberish!)

Joint Publication 5-0, Joint Operations Planning, directs that commanders conduct center of gravity analysis, both from the friendly and enemy perspective, at both the strategic and operational levels and state their findings in Department of Defense war plans.[20]

If our cathartic reaction to defeat in Vietnam was the genesis on neo-Clausewitzian thought, our debacle in Iraq may be its swan song. There is a stunning inability of modern American generals, no matter how intelligent, dedicated, and well trained in their craft, to identify and defeat the enemy center of gravity and thereby achieve rapid and unquestioned victory in Iraq. In the past few years virtually every senior general officer involved in the initial phases of Operation Iraqi Freedom has confessed that he misjudged the ever-shifting Iraqi center of gravity. High-ranking officers of appropriate intelligence cannot, by their own admission, identify centers of gravity, much less attack them to achieve the promised decisive victory. Still, institutionally the army is convinced that a center of gravity must exist because . . . Clausewitz said so! And this belief persists despite the fact that we cannot agree about the nature of center of gravity and how to address it, which is precisely why the army's current doctrine regarding center of gravity is so confused and muddled.

A recent article entitled "Understanding the Link between Center of Gravity and Mission Accomplishment," written by two members of the Army War College faculty and published in *Military Review*, the professional journal of the U.S. Army, concludes:

> Clausewitz made clear the link between neutralizing the COG and victory, and U.S. joint doctrine has adopted Clausewitz's COG concept. The emerging joint doctrine on operations and planning must take the next step: it must clarify the link between neutralizing the COG and accomplishing the mission.[21]

Army doctrine embraces as its centerpiece of strategic analysis a concept that for all practical purposes is indefinable and not particularly useful in winning our nation's wars. Rather than invest more lives in a misguided effort to make the writings of a nineteenth-century German philosopher work for America in the twenty-first century, we should ask if there is perhaps a better point of departure for our doctrine. I argue that there is.

Chapter 2

A Framework for Understanding America's Wars

My criticism of Clausewitz is that he addresses an archaic form of warfare, wars between kings, that has been swept away by the democratic demands, whatever their forms, of peoples worldwide. The key question in modern warfare, especially for America, is not how to destroy enemy armies but rather how to defeat enemy governments and then establish better governance for their populations. Modern wars are not primarily about armies and battles, they are about populations and governments. They are not about winning limited wars for temporary advantage; they are about establishing a permanent peace in which all can flourish. The U.S. Army, shaped as it was by American democratic values, once understood this. We would be wise to relearn it.

My method is to first lay out the theory, much like an attorney would make his opening argument. I state what the facts will show, then guide you through American military history to link the facts to the theory. As in statistical analysis, the best theory is the one that best fits the facts; but without the theory, one doesn't know what facts are important. What I hope will become clear is that governance, not center of gravity, is a far more useful framework for modern war doctrine. Not all of the data will be pleasant. War is about killing, often extreme killing, but it is far better to recognize this reality than to march off to war unaware of its true demands.

Later I briefly discuss modern wars not involving America to see if the pattern is generally true or peculiar to America. Then we look at our wars in Afghanistan and Iraq. Once confirming the general applicability of the model, I discuss its implications for American military policy.

My thesis is that wars are defined by their strategic aim and generally fall into two significantly different categories: offensive and defensive. This distinction, however intuitive to most people, is not expressed in our current military doctrine. Wars can be further categorized with respect to the governance of populations. The attitude of the indigenous population toward the existing or proposed governance determines the nature of the war. Wars of occupation occur when foreign powers seek to force new regimes of governance on populations that are at least partially resistant. These wars are always offensive in nature. Wars of revolution are internal struggles where the revolutionary aim is to forcefully establish a new form of popular governance. Revolutionary wars are considered to be offensive in nature because they seek to disempower the existing order and establish a new regime of governance. Defensive wars seek to preserve popularly accepted forms of governance from external aggression or revolution. Wars of liberation seek to restore popular systems of governance that existed prior to invasions by other countries or undemocratic coup d'états. These wars are often part of a counteroffensive waged by an alliance of defending nations.

Limited wars, the focus of Clausewitz's writings, seek to redefine behaviors and establish new international agreements between nations, but they do not directly challenge domestic governance arrangements or national survival of either warring party; these wars are ended once the specific issues at hand are resolved. From the perspectives of the nations involved, limited-purpose conflicts can be viewed as either offensive or defensive in nature; indeed, there is often disagreement on that matter. The important point is that governance doesn't change.

As obvious as it may sound, in all wars defeat of the enemy military is often a necessary condition for successful conflict termination. In offensive wars, defeat of the enemy government and its popular base is also required, as is the imposition of military government to force political reconstruction and provide some measure of economic assistance. In wars of liberation, the army must support the restored government until normalcy returns.

Victory in offensive wars, especially where one nation seeks to impose a new form of governance on another nation, is exceedingly difficult to achieve, as history repeatedly demonstrates. Imposing by force of arms a new governing system on a resistant population is not an enterprise for the faint of heart. No matter how virtuous the aggressor may feel his cause to be, the defending population, or at least some segments of it, must be assumed to feel as strongly toward its traditional form of governance. Dictators, warlords, tribal chiefs, religious zealots, and other nondemocratic leaders repeatedly demonstrate their abilities to call their populations to the defense of their political systems and compel sacrifices that we in modern America can scarcely imagine.

The requirements for victory in wars of occupation generally follow a chronology: defeating the enemy armed forces, occupying enemy territory and population centers, breaking the enemy population's will to further resist, disestablishing the defender's government, establishing the occupier's monopoly of force (military and police) over the population, forming a new national government, establishing popular loyalty to the new government, investing in the new indigenous government police and military responsibilities, and ultimately reducing or withdrawing the occupation force. But often in irregular wars, as we'll see later, the main fighting occurs after the new government is imposed, and the phases of the war of occupation are sequenced differently. The entire process of imposing new governance could take ten to fifteen years.

The attacker must recognize that a large amount of killing will probably be required to make the attacked population concede to occupation and new governance and be prepared, both politically and militarily, for the task. Few societies capitulate while they still have the demographic means to resist. Indeed, looming demographic collapse from combinations of death, disease, and starvation seems to be the proximate cause of most surrenders to occupying armies.

Based on data from modern wars, death rates of at least 5 percent and as much as 20 percent may have to be inflicted on the defending population before it will accede to the occupier's new governance. Generally, most of these casualties will be military-age males, among whom casualty rates—killed, wounded, captured—will likely exceed 50 percent before the society loses its demographic ability and political will to further resist. As families lose their sole surviving sons, peace becomes the only way to carry their legacy forward.

It is also regrettable that the attacker will have to demonstrate his willingness to kill the defender's women and children, often by killing 1 percent or more of the civilian population, to ensure that defending leaders see clearly that their only option to extermination is surrender. It is also the case that massive casualties, property destruction, and near-starvation economic conditions are the only proven means to convince the defending population that its wartime leaders have betrayed their interests, lied to them about the nature of the war, and no longer deserve their respect or obedience. Indeed, the devastating traumas of war are a necessary antecedent to the identity shift and political rebirth that the attacker seeks in resorting to offensive war. Psychologically and practically, the complete collapse of the old regime is required for the birth of the new.

There is no indication that defending populations have lowered their demographic loss tolerance in recent decades, or to suggest that modern offensive wars can be won cheaply or quickly. If Americans have become squeamish about war casualties, we must realize that many of our potential enemies have not.

Almost inevitably, wars of occupation become wars of attrition. The numerous battles and engagements required to achieve the necessary demographic effects smooth out the incidental effects of genius, superior generalship, luck, and surprise. As Clausewitz observed, genius and chance often determine the outcomes of individual battles. But they do not measurably affect the outcomes of wars of attrition. Invariably, the military decision becomes a function of the average war-fighting ability of the two competing sides. Cultural, technological, economic, demographic, and political factors are the main contributors to

Taxonomy of American Wars and Conflicts from the American Perspective

Occupation (Offense)	Revolution (U.S. Support of; Offense)	Defense	Liberation (Counteroffense Phase of Defense)	Limited Purpose
Indian Wars*	Cuba (1898)	Revolutionary War	World War II (1943–1945)	Barbary War
War of 1812 (vs Canada)	Provide Comfort/	Civil War (South)	Bay of Pigs	War of 1812 (vs British)
Civil War (North)	Northern Watch	World War I	Afghanistan (1979–1989)	Mexican War
Cuba (1899–1902)		World War II (1941–1942)	Desert Storm	Spanish-American War
Philippine War		Cold War	Urgent Fury	Punitive Expedition
World War II (1945)		Taiwan (1949–present)	Just Cause	Cuban Missile Crisis
Korea†		Korea‡		Desert 1
Vietnam		Former Yugoslavia		Afghanistan (1989–2001)
Restore Hope		Korea‡		Iraq (1990–2003)
OEF (Afghanistan)				
OIF (Iraq)				

* Aim was removal   † North of 38th Parallel   ‡ South of 38th Parallel

Requirements for Victory

	Defeat of Enemy Military	Defeat of Enemy Government	Defeat of Enemy Population	Political Reconstruction	Economic Assistance
War of Occupation	Required	Required	Required	Required	Later phases
Revolutionary War	Required	Required	Required	Required	Required
Defensive War	Whichever comes first	Whichever comes first	Not required	None	As required
War of Liberation	Whichever comes first	Whichever comes first	Not required	Minimal assistance	As required
Limited War	Insofar as necessary	Insofar as necessary	Not required	None	None

Demographic Data from Selected Successful Offensive Wars of Occupation[1]

	Defender	Defender Population	Defender Military/Civilian Deaths	Percentage Military Casualties*	Result of War
King Philip's War (1676)†	New England Indians	20,000	3,000 (15%)	Unknown	Surrender/Removal
Civil War	Southern states	5,600,000	300,000 (5%)	60%	Surrender/Reconstruction
Philippine Insurrection	Filipinos	7,000,000	220,000 (3%)	Unknown‡	Surrender/Colonization
2nd Boer War	Boers	200,000	37,000 (18%)	90%	Surrender to British
World War II (1943–1945)	Germany	70,000,000	4,500,000 (7%)	90%	Unconditional surrender
World War II (1943–1945)	Japan	70,000,000	3,300,000 (5%)	30%§	Unconditional surrender

* Dead, WIA, POW, and MIA in male military-age population.
† First large Indian war in New England; defined the pattern for the next 225 years.
‡ Standard estimates are 20,000 killed out of 100,000 who fought, plus 200,000 civilian deaths.
§ In August 1945 the bulk of the remaining Japanese army was isolated and trapped on the Asian mainland and South Pacific islands, but not yet POW.

Demographic Data from Selected Failed Offensive Wars of Occupation[1]

	Defender	Defender Population	Defender Military/Civilian Deaths	Percentage Military Casualties*	Result of War
King Philip's War (1676)	New England Colonists	40,000	1,000 (2.5%)	Unknown (probably about 10%)	Colonist counteroffensive against Indians (1676)
American Revolution	Colonial Patriots	1,500,000+ (out of 2,500,000)	25,000+ (2%)	10% (deaths)†	British failure/Independence
USSR World War II	USSR	196,000,000	16,000,000 to 26,000,000 (8–13%)	75% total; 35% KIA, POW, MIA	German defeat/Soviet counteroffensive
Vietnam War (1961–1975)	Vietnamese Communists	43,000,000 (1970) All Vietnam	2,000,000+ (5%)	25% deaths; WIA unknown‡	Vietnamese reunification
Soviet-Afghan War	Afghanis	15,500,000	1,300,000 (8.4%)	Unknown	Soviet failure

* Dead, WIA, POW, and MIA in male military-age population.
† Patriot deaths were 25,000 of the 250,000 who served. 250,000 would represent full mobilization for a Patriot population of 1,500,000. 30,000–50,000 Americans fought for England.
‡ VC and NVA deaths were 1,000,000 out of the 4,000,000 who fought. ARVN, which reached peak mobilization of over 1,000,000 in 1972, lost 250,000 dead and 784,000 WIA.

war-fighting ability, and these factors are unlikely to change significantly in less than half a generation, the span of most wars. Loss exchange ratios (that is, casualties inflicted divided by casualties suffered) generally reflect the relative average war-fighting abilities of the combatants.

Attrition warfare for the attacker means sustaining sufficient forces in the field to inflict the necessary casualties on the defending military-age male population (about 50 percent casualties) and the overall population (about 5 percent killed) to reach the demographic effects necessary for the defender's capitulation. For the defender, who generally is on the wrong side of the loss exchange ratio, attrition warfare means inflicting enough casualties on the attacker to break his political will. United States Army doctrine up through the Vietnam War was fairly consistent in its viewpoint that superior American firepower, provided by a superior economic and technological base, is the best way to inflict high casualties on the enemy and preserve friendly forces—that is, optimize loss exchange ratios.[2] The military draft, a fixture of twentieth-century U.S. military policy until the 1970s, provided the other necessary component, manpower, for the army's attrition strategy.

In the case of modern peer societies, which can efficiently mobilize and deploy their male populations for conventional defense, military and civilian capitulation may be simultaneous. Examples are the Confederacy and Germany and Japan in World War II. But in many cases, the enemy will not efficiently mobilize its defensive resources, and its available military will be defeated well before the population is prepared to accept conquest and imposed governance. In these instances, the defender commonly turns to protracted guerilla war against the occupying military and the unpopular imposed government. In these cases, hard-war counterinsurgency strategies will be required to achieve the defeat of the defender's male population, the destruction of the guerilla's economic support, and the surrender of the popular will to further resist. These irregular defensive wars, like conventional wars, seem to escalate into wars of attrition and follow the same demographic rules, continuing until either the attacker grows tired of the war or the defender reaches the point of demographic inability to further pursue the war. There may be some societies that will resist occupation almost to the point of extermination, especially if they feel that extermination is the true motive of the attacker.

Importantly, it is the defending population, not the attacker, that determines when the war against the ardent occupier is over. The defender, if he elects, can move from a strategy of annihilation (of the invading army in battle), to a war of attrition, to a protracted guerilla struggle, to a terrorist campaign, to civil disobedience, and ultimately to indifference and idleness if needed to rob the attacker of meaningful victory and increase his cost of occupation. The evidence shows that defending populations are resourceful and resilient in their means and methods of resistance. In essence, the occupier is demanding that the occupied population assume a more agreeable collective identity. The degree to which the defender resents that newly imposed identity, which inherently entails some degree of subservience, determines the strength and length of resistance. Indeed, resentment against Soviet occupation led to decades of resistance in Central and Eastern Europe. A number of anticolonial wars lasted a century or more until the colonial occupiers finally agreed to leave. Some of the identities that occupiers offer their new subjects are simply too odious to ever gain popular acceptance.

All successful occupation armies, then, are essentially agents of revolution in the countries they occupy, aspiring to bring about governmental and societal changes that will transform the defeated nation and inure it to a new identity that will ensure alliance and comity going forward. Great care must be taken in crafting the new identity and selling it to the target audience. The soldiers of the occupation army must see themselves as the agents and emissaries of revolutionary change and conduct themselves accordingly, realizing that the occupied population will judge in some measure the sincerity and desirability of the occupier's proffered new identity and governance by the conduct of the occupying soldiers. Successful revolutionaries, from George Washington to Mao Tse-Tung, insisted that their troops deport themselves with discipline and generosity, knowing that the people would never rally behind causes they perceived to be unjust or self-serving. The occupations of most European colonial regimes, Nazi Germany, and the Soviet Union largely failed because the target populations rejected the identity imposed by the occupier, based on the terror tactics of the occupying troops, the lack of satisfactory justice for the population, and the reality of second-class status. Occupations that do not inspire viable new identities and the companion social revolutions in the target population ultimately fail.

Creating a New Government in the Occupied Country

A primary duty of an occupying army is to establish military government in the areas under his control.[3] Having rendered the previous political regime null and void, the security and welfare of the population are now a military responsibility. The army must train and have ready sufficient military government units to provide the necessary governance. These units should be organized for all levels of government, national through local, to perform all executive and judicial branch functions, including, especially, public safety, justice, court, public works, public health, education, and economic redevelopment. The military government must be supported by adequate military units and personnel: police, constabulary, civil engineering, legal, transportation, and quartermaster, as well as judges, lawyers, cultural experts, and linguists. Importantly, these military governance units should be under separate command from the combat units that are tasked to defeat enemy combatants, not provide governance to the occupied populations. It is impractical for commanders to perform both missions at the same time.

Establishing the rule of law, even if martial law, is perhaps the most essential operation of military government. The citizens must be told immediately the purpose and duration of the occupation, what is expected of them, and the penalties for noncompliance. To the maximum extent possible, government workers below the political level must be co-opted to work under the supervision of the occupier to restore or maintain accustomed public services. The population must be documented and its activities controlled. No police, militia, or military force can be allowed that isn't strictly controlled by the occupier. It is of paramount importance that the military governance be seen as firm, humane, competent, and just.

Occupation is a continuation of attrition warfare, where the attacker trades personnel, time, and money to win more acceptable governance of a once hostile population. High concentrations of military governance and security personnel, perhaps ten soldiers per thousand of population, will be needed to perform all the military government functions.[4] These forces are in addition to those still battling enemy combatants. This means that the creation and deployment of the military government forces cannot be an afterthought, not merely a simple retasking of units no longer needed in the fight. Defeating the defender and military governance are simultaneous and competing elements of the overall plan, each vying for scarce resources such as manpower, strategic lift, and in-theater logistic support. Both are critical to overall mission accomplishment: winning the war and the peace.

As the war winds down, combat units can be re-formed and retrained as constabulary units to assist in security functions under the direction of the military government. Eventually, as relations between the occupier and occupied normalize, more governance work can be turned over to local citizens. But at the climax of the war, the combined force structure bill for separate combat and military governance operations will be high. Still, there is no avoiding that bill if one wishes to achieve a useful peace.

It must also be noted that military governance is not counterinsurgency; it is an operation designed to prevent an insurgency from developing by controlling the population and meeting its essential needs, including security.

Of course, insurgents are mainly military-age men. The more of these men who were killed in battle prior to the occupation, the fewer who will be able to assume the insurgent role afterward. Those who are dead can no longer resist. There is a perverse dynamic here: a quick, relatively bloodless invasion provides more fuel for insurgency and a more problematic occupation, while a bloody conventional war removes the fuel for insurgency and makes the occupation easier. The military government must pay special attention to the remaining young men, whose support will be crucial to success.

Additionally, although it is the responsibility of military government units to provide police security and, in doing so, police security units will inevitably become involved in investigations of insurgent activities, military governance police units should not become involved in counterinsurgent combat operations. It is impossible to fight a population and govern it at the same time.

Formation of a new indigenous government acceptable to the occupier is always problematic. Because the new government must reflect, bureaucratically and procedurally, the new national identity that the occupier wishes to impose, acceptable indigenous popular government may be unachievable. Much rests with the new population accepting the offered new identity. The political, cultural, and economic leaders from the defending populace will have decidedly mixed feelings toward their new masters and will tend to resist new governmental processes, which invariably reduce their traditional power and established social norms. Those publicly supporting the occupiers will likely be seen as collaborators and treated harshly by the populace. Returning émigrés, who have to whatever degree accepted the occupier's values, may be viewed as outsiders, traitors, or usurpers. Given the variety of opposition, no new political class will coalesce unless and until the occupier can ensure its safety through its monopoly of violence.

There are also limits to the reforms that an occupier can reasonably achieve. The cultural, religious, economic, and clan relationships built over centuries cannot all be remade in the limited time and with the limited resources available to the military government. Fundamental change of that great magnitude takes generations, and the military can only play a small part in the conversation that changes directions of civilizations. Rather, the occupier must be pragmatic. The first goal is to create a permanent ally, next to provide for popular government, and only lastly to re-form the society along American lines. Building better governance upon traditional foundations is more likely to succeed. An occupying power trying to impose a democratic government on a resistant population can expect nearly insurmountable problems. In the few instances of success—the reconstructed South, post–World War II Germany and Japan—the defeated populations had had recent experience with representative government.[5] There is no empirical evidence suggesting that representative government can be imposed within a matter of years in countries without democratic traditions.

Americans have traditionally inaugurated programs of economic reconstruction and economic aid to occupied populations to promote goodwill, good governance, and human decency. Economic attraction has always accompanied military chastisement since the latter stages of the Indian Wars through the war in Iraq. At the tipping point, economic benefits may sway the decisions of defenders to cease resistance. However, resistance movements rarely quit fighting as long as they still have the demographic ability to continue the war and see reasonable opportunities for political or military success. Often, resistance groups will find ways to take advantage of economic aid even while continuing their resistance. Economic aid in all its forms is never decisive, but it is certainly useful in obtaining agreements to end resistance once the population is already leaning in that direction. Raising friendly and fence-sitting populations from the economic nadir often created by war is a powerful tool in cementing continued friendship.

Restoring sovereignty to the occupied population is best done in stages. Supervised by the occupier, local government may be restored after the proper vetting of officials. Nevertheless, the occupier, as the supreme executive authority, must be prepared to remove uncooperative officials and veto unhelpful statutes and policies. Eventually, elected assemblies must be allowed at increasingly higher levels. Initially, however, they should be advisory in nature with the military government retaining all executive authority. Again with occupier oversight, lower courts can be returned to the indigenous judiciary. Indigenous military forces may be organized and equipped, but they must be placed firmly under the occupier's command and control. If, over a period of years, a true partnership develops between the occupier and the occupied, then full sovereignty should be restored to the occupied nation. This includes all executive, legislative, and judicial authority, including command of its military, the only qualification being that its military be bound by alliance and other considerations to the former occupier. Done for the mutual benefit of all, the occupation process can result in abiding common interest and indeed friendship between the former foes. Selfishly done, the war and occupation may lead to centuries-long enmity, committing future generations to additional wars.

Need to Manage Unintended Consequences

Failed or improperly terminated wars of occupation create power vacuums and instability that may unleash powerful revolutionary forces. Revolutionaries exploit the chaos of war, the overall dilution of the military strength opposing the revolutionaries, the anarchy of transitions, and ungoverned no-man's-lands to further their own agendas.

Another way forward all too often emerges. For example, the Communist revolution in Russia was an unintended consequence of failed Russian and German imperial warfare in World War I; the rise of Red China was enabled by the chaos resulting from the Japanese invasion; and the successful anticolonialist wars of liberation that followed World War II were a direct result of postwar Europe's military and economic exhaustion and consequent inability to police their prewar empires.[6]

Iraq is only the most current example of a nation unwittingly unleashing radical forces by an ill-considered military offensive. The moral is that an attacking nation should not attempt to seize more than it can realistically control. Nor should a nation defeat an enemy government without consideration for the forces that it keeps under control. Revolution is a too likely result. The need to control conquered populations is absolute.

Defensive War

Victory in defensive war, however difficult it may be to achieve, is conceptually simple compared to offensive war. Defensive wars are terminated by the defeat of the attacker's military in the field or the defeat of the attacking government's political will to continue its offensive war of occupation. The basic task of the defender is to make the military or political cost of war too much for the attacking government to continue to bear. Often the attacker's cost threshold is surprisingly low. Deterrence is a form of defensive strategy designed to make the anticipated cost of any aggression too great for the nation considering an offensive war. Political reconstruction after a successful defense is either unnecessary or minimal, because the defending government either has remained in power or can be easily reconstituted in liberated areas. Return to the political status quo antebellum in liberated areas is seldom a challenge for the successful defender, assuming that the government enjoyed broad public support prior to the war. Economic assistance to citizens harmed in the defensive struggle may have to be provided consistent with the traditional practices of the victorious government.

Limited War

Limited wars are wars between nations in which neither side wishes to change the government of the other but rather change a nation's policies, alliances, spheres of influence, or possession of border areas or other territories. These wars are the focus of Clausewitzian thought. Rationally, these wars demand a cost-benefit-risk analysis from each party in the conflict: is the war worth it or not? However, as Clausewitz observed, the passions unleashed by the saber rattling and initial bloodshed may sweep away the careful calculations of elite cabals and compel an unintended general war, an ever-growing danger as people worldwide increasingly view themselves as democratic participants, rather than pawns, in the affairs of state. To keep a war limited, both sides must succeed in clearly defining the limits of the war and maintaining objectivity—both difficult tasks. Often, third nations or international bodies offer to mediate ends to these wars, the two nations involved being too embittered to make peace on their own. The peace, when it comes, may not assuage the animosities that led to the war or developed from the war. Some societies have long memories, and an enemy once made might become an enemy forever.

I have deliberately gone into great depth in the above discussion only with respect to offensive warfare. The sections above on defensive and limited war are short because, conceptually, victory in these wars is fairly straightforward. Destruction of threatening enemy forces in battle or by long-range strikes is not a particularly difficult task for the U.S. armed forces today. From a purely military standpoint, the United States has already solved most of the problems posed by these forms of warfare, which dominated our recent Cold War experience and still are the centerpiece of our post-Cold War force structure and training. There is little that needs to be added here, certainly nothing that would merit a rewriting of doctrine, other than to repeatedly warn against escalating limited wars into far larger wars that are not consonant with our interests.

The prosecution of revolutionary war has not been discussed in detail because this is not a problem that the American military needs to solve. We are not revolutionaries; we are the defenders of a U.S. government that has no domestic insurgent challenge. Our wars are beyond our borders in other peoples' countries.

Supporting defensive counterinsurgency efforts of friendly and allied governments overseas, however, is a crucial U.S. military mission, but again one that the U.S. Army seems to understand institutionally and at which it is generally successful.

Supporting overseas insurgencies is already a recognized army mission, one that we executed well in support of the Afghan resistance to Soviet occupation. The morphing of that resistance into the Taliban and al-Qaida might cause us to consider whether we can actually control the revolutions we foment.

My focus is on offensive war where we chose to invade another country, disestablish its existing government, and replace it with a new government that will support our interests. This is because offensive war is the mission we have most recently embarked upon and cannot seem to, at least lately, get right. I discuss the other forms of war only to clarify what I mean by offensive war and to isolate the phenomenon for analysis and conclusions.

Chapter 3

America's Military Legacy through World War II: Case Studies in Successful Offensive War

I promised that the strategic military doctrine expounded previously would not be theoretical or speculative but rather empirical—that is, based on observations of past and current wars. The data shows that patterns do exist, patterns that should form the basis of doctrine. Clausewitzian dictums extolling the attack on the center of gravity and the desire for decisive battles are not supported by the evidence. Rather, attrition warfare seems to be the strategy most likely to bring offensive wars to a successful conclusion. Further, combat success alone is insufficient in winning offensive wars; only properly executed military governance, based on the rule of law, can consolidate the battlefield victories into a lasting and favorable peace.

Analysis at the strategic level shows that the best predictors of success or failure in any war are found in the broad outlines of the war. The three key factors are the envisioned post-conflict governance; the willingness and ability to win a war of attrition; and, if prosecuting an offensive war, the willingness and ability to impose new governance on the occupied population. All else is detail, important only insofar as it contributes to one or more of the three key factors.

The Indian Wars

The wars against the Indians lasted nearly three centuries and are the formative American military experience. The American understanding of democratic warfare, attrition, and imposition of a favorable permanent peace begin in these most savage and seminal wars.

For the better part of two centuries, the Indian wars were fought mainly by militias composed of the armed citizenry living along the frontier. The militia laws in the American colonies generally required every able-bodied man of military age to maintain a firearm and to drill, once a month or so, with his militia company, often the other men living in his county. The colonial governor and other more local officials could call out the militia in the event of war with the Indians, among other less-pressing reasons.

Militias were a practical solution for colonial warfare. The small colonial agrarian populations could not afford standing armies or prolonged absence from their farming, hunting, and other life-sustaining pursuits. Neither could the Indians. If permanent peace between the Indians and the ever-growing English population was impossible, so was permanent war. So the Indian wars were mainly episodic, the Indians attacking to drive back the encroaching white settlers, the militias responding with a counteroffensive designed to remove the Indians permanently from the area. There was no American attempt at Indian governance until the nineteenth century. Rather, removal became the objective, and the Indians recognized this reality and responded as best they could.

The early Indian wars were the most savage ever fought by Americans, testing the limits of mobilization and attrition. The Good Friday Jamestown Massacre of 1622 is among the worst days of warfare imaginable. On that day the Powhatan Confederation, which consisted of about ten thousand Indians living in two hundred villages, surprised and killed 347 of Jamestown's mostly male colonists, almost one third of the population.[1] The Indian destruction of the colonists' crops led to another 500 deaths by malnutrition and disease over the next winter. Only continued English immigration, often by people unaware of what they were getting into, allowed the colony to rebuild; the population again reached over 1,200 by 1624. Martial law was declared immediately after the Indian attack, and the Jamestown militia waged an unrelenting war of attrition, known as the First Tidewater War (1622–1632), against the Indians, with the main tactic being the burning of nearby Powhatan settlements and crops.

Beginning in 1629, the colonists resolved to remove all the Indians from the areas east of Jamestown and build fortifications to the west.[2] The scheme succeeded only temporarily. In 1644 the Powhatans again launched surprise attacks on the Englishmen, killing nearly five hundred of the approximately eight thousand settlers on the first morning of what would become known as the Second Tidewater War (1644–1646). Again the militia responded by attacking Indian raiding parties and nearby villages and crops. Additionally, one of every fifteen able-bodied men was enlisted to serve in a standing military force for a period of one year, which allowed the colonists to strike deep into Indian territory and establish fortifications up to the fall line of the James River. Over time, the weight of the colonists' military effort was too much for the Indians to bear, and again they had no choice but to cede land to the English in return for peace.[3] Indian death tolls in the Tidewater wars are not known, but the demographic collapse of the region's Indian population from war, disease, starvation, and economic calamity coincides with these wars.[4]

King Philip's War (1675–1676) was New England's first great Indian war, but it followed the same pattern as the wars in Virginia. About seven thousand of the twenty thousand Indians living in New England launched widespread surprise attacks on the forty thousand European settlers intermixed among the Indian populations. Indian warriors raided and burned outlying settlements, massacring hundreds. Entire regions were abandoned, some colonists feared permanently, as terror-stricken frontiersmen retreated to more populated and protected areas. Colonial militias rallied, however, and began the counteroffensive that would kill perhaps three thousand Indians and destroy numerous Indian villages until, finally, their chiefs agreed to surrender their lands and forswear future warfare. The war featured savage winter offensives by both sides aimed at destroying settlements and populations when they were most vulnerable, a favored tactic in the Indian wars for the next two hundred years.

The war is illustrative of the ability of normal populations to withstand attrition. The attacking Indians failed to inflict critical levels of casualties in their initial assault. About a thousand settlers—2.5 percent of the total—died in King Philip's War. The number of deaths was enough to raise passions for revenge but not enough to cause surrender. The colonist counterattack, in contrast, killed 15 percent of the Indian population—men, women, and children—placing the very survival of the Indian population at risk. Indian tribal chiefs realized that peace at any price was preferable to continued war and extinction, and they would have to accept removal to survive.[5]

Over the next two hundred years, similar American tactics forced countless Indian chiefs to cede to America's expansionist demands. As the Anglo-American populations grew and pressed their strategic offensive, the defending Indians were pushed ever westward. Eventually the white populations along the Atlantic coast were no longer in contact with the Indians, so militias from western territories and states, combined with regular army forces, took up the battle. The Indian wars reached their zenith in the War of 1812, during which the major hostile tribes remaining east of the Mississippi were soundly defeated. The successful American generals rode their victories to the White House: William Harrison, the victor over Shawnee leader Tecumseh's confederation at the Battle of the Thames, and Andrew Jackson, who led the army that routed the Indians of the Mississippi Territory in the Battle of Horseshoe Bend.[6] It was during Jackson's presidency that the Indian Removal Act of 1830 finally codified America's two-hundred-year-old removal policy by forcing the Indians who would not agree to live on reservations out onto the then relatively unpopulated Great Plains.[7]

It was in the Trans-Mississippi, after the Civil War, that the Indian Wars reached their denouement. Operationally, the later Indian wars were little different from the first, featuring winter campaigns and converging columns, all designed to attack the Indians' villages and destroy their very means of survival. The tactic forced Indian males to fight under the most disadvantageous conditions at a time when the growing wealth and technology of the United States allowed the army to keep up what amounted to a permanent offensive.[8]

The real innovations of the later Indian wars were the creation of the Bureau of Indian Affairs, founded in 1824 in the Department of War and transferred to the Department of the Interior in 1849, and the reservation system for permanent governance of the Indian tribes. The general idea was that the army would compel the Indians to agree to live on the reservations while the Bureau of Indian Affairs would administer the reservations in accordance with a perhaps dubious notion of retained Indian sovereignty.

This division of labor was always contentious in the army, which institutionally prefers unity of command. The army argued that its good work at compelling the Indians to move to the reservations was ever being subverted by what it viewed as the manifest inadequacies of the Indian bureau in delivering the promised benefits of reservation life.

Indeed, army officers and units were often assigned duty with the bureau when the civilian agency fell short. But the government believed that the soldier who fights the Indian could not be the soldier who governs and civilizes the Indian, and for that reason it kept the two functions as separate as possible. Over the decades, the army came to see that hard-won victories and pacification campaigns could only be secured by firm but fair paternalistic government aimed at the eventual economic and political transformation of the vanquished.[9]

These wars in the Trans-Mississippi followed a predictable pattern. Indians would initially refuse to go to their designated reservation or later leave the reservation for their traditional lands. The army would chase down the tribe and deliver an ultimatum, which the Indians would refuse. Superior Indian mobility and ability to live off the land offered an initial advantage, but the army would pursue relentlessly. The constant skirmishes against heavily armed army forces sapped Indian fighting strength, and the need to constantly move Indian villages caused hardship and malnutrition among the women, children, and elderly.

Even the most warlike of chiefs would eventually realize that the survival of his tribe required him to swallow his pride and accept the inevitable. In one of the final Indian campaigns, the Nez Perce tribe, which numbered eight hundred people of whom three hundred were warriors, refused to move to the reservation as the army insisted. Instead, Chief Joseph led them on a heroic yearlong, 1,600-mile fighting retreat across the continental divide. Having seen half his people die, reduced to only three hundred women and children and eighty-seven men, Chief Joseph stated, more eloquently than I ever could, why nations losing a war of attrition eventually have to surrender.[10]

> Tell General Howard I know his heart. What he told me before—I have it in my heart. I am tired of fighting. Too-hul-hul-sit is dead. Looking Glass is dead. He-who-led-the-young-men-in-battle is dead. The chiefs are all dead. It is the young men now who say "yes" or "no." My little daughter has run away upon the prairie. I do not know where to find her—perhaps I shall find her too among the dead. It is cold and we have no fire; no blankets. Our little children are crying for food but we have none to give. Hear me, my chiefs. From where the sun now stands, Joseph will fight no more forever.[11]

Only the cruelty of attrition warfare could reconcile the Nez Perce to their new identity. No longer were they a nomadic people living on their ancestral lands in their traditional ways; their very survival required them to accept the reduced status of the reservation and the governance of the Indian bureau.

England's Offensive Failure in the American Revolution

The next war for consideration is the American Revolution, a war that Clausewitz did not analyze as he developed his theory extolling decisive battles at centers of gravity. The demographic data from the conflict reinforces the need for inflicting widespread casualties in offensive war.

The Revolutionary War is misnamed. It was not the Americans who sought revolution but rather the British who sought to impose on the colonies new governance: direct rule from London. The colonies initially sought only to retain their previous forms of democratic self-government based on their colonial assemblies and legislatures, but a year of English aggression caused the colonies to unite in nationhood in favor of independence.

The American political leadership in place before the war was largely sustained by Patriot forces throughout the conflict and empowered by the people after the war as the founding fathers of the new nation. Revolutions don't sustain the preexisting political class; rather they remove it. This was no revolution; this was a defense, and it was understood to be so at the time.

British military occupation of Boston, the first permanent stationing of Redcoats in the American colonies, put all Americans on notice that their freedoms were at risk and sparked resistance throughout the colonies. The English declaration of military rule throughout Massachusetts in 1774 caused the Massachusetts militias to organize for common defense and inspired the

First Continental Congress, the first effort to unite all the colonies in the cause of traditional liberties.

The British attacks at Lexington and Concord, and their defeats on those battlegrounds, led to Patriot takeover of virtually every governmental and civic organization throughout the colonies and, shortly thereafter, the Second Continental Congress, the formation of the Continental Army to fight the British, and the 1776 Declaration of Independence. A well-established militia system, the people in arms supporting their democratically elected officials, delivered the colonies efficiently and permanently (some wavering aside) to the Patriot cause. After 1775 the task for the British was to reverse that decision.

For their part the British could not bring themselves to believe that the majority of Americans, at least the ones who counted, preferred independence to rule by the Crown. Although Clausewitz had not yet written of center of gravity and war, a mindset developed among the British that if what Clausewitz would have considered the Patriots' center of gravity could be broken, then the colonies would fall into line.

The British selected Boston as the Patriots' first center of gravity, but its seizure served only to unite the Americans in common defense. Forced by Gen. George Washington to retreat from Boston, the British regrouped and attacked the next selected center of gravity, New York, and scattered most of Washington's Continental Army. The response was a Patriot strategy of protracted struggle, stealing victories where they could, while the British paraded from one hoped-for center of gravity to another—Philadelphia, the Lake Champlain-Hudson River corridor, Charleston, and eventually Tidewater Virginia—in an eighteenth-century version of whack-a-mole.

Despite the many British offensive victories, the Continental Army always reemerged, often stronger than before. And even though the British army often advanced from place to place, rarely were the British able to more than temporarily disrupt the virtually universal population control maintained by the Patriot militias and civil authorities. British forces in America reached a peak strength of thirty thousand in 1777, enough to defend its base at New York and on and off occupy a few other coastal cities but never enough to control the vast countryside and the myriad cities and towns scattered across America. Neither could the British force a war-ending defeat on the Continental Army, which by November 1778 had reached a strength of thirty-five thousand men. From Concord to Yorktown the fighting lasted over six years and included over 1,300 separate land engagements, none of which was decisive. Eventually, following their defeat at Yorktown, the British quit the war due to its ruinous expense, its lack of reasonable prospects for success, and its unpopularity in the home country. Even in England the Patriots' defense of democratic liberty against the king's assault found broad support in the high tide of the Age of Enlightenment.[12]

Historians estimate Patriot deaths from all causes at 25,000 out of 250,000 who served in the war. (Most were militia.) This participation rate suggests a Patriot population of perhaps 1.5 to 2 million of the 2.5 million colonists. (About 30,000 to 50,000 Americans fought for the Crown, suggesting a Tory population of perhaps half a million. The Tories suffered heavily in battle for their treachery, as Kings Mountain bears witness, suffering about 3,000 battle deaths.[13] At war's end, 80,000 to 100,000 Tory citizens chose to abandon their homes and flee with the retreating English army, eventually leaving America behind.[14])

Patriot deaths were no more than 2 percent of the revolutionary population, well below the casualty rates of the Indian wars and well below the demographic threshold that would have brought victory to the British. Fortunately for us, the British did not target the civilian population, a restrained approach that, however high-minded, played into the Patriot protracted war strategy and allowed the Americans to avoid sacrificial combat on unfavorable terms. Ultimately, England hadn't the collective passion to prosecute the harsh attrition war necessary to impose rule in the colonies and so would never seriously test the limits of American commitment to continued independence. As a result, the attrition war favored the Patriot cause. In essence, the British failed because they tried to fight a dynastic offensive war in a new democratic age.

It must be noted that the American forces in the thirteen colonies operated under the strictest codes of conduct. General Washington sensed that indiscipline by his Continental Army would endanger popular support for the new Congress in Philadelphia and force Americans to the British and their Tory allies for protection. Washington knew that the civilians whose sentiments could be swayed largely judged the worth of the new government through the actions of its soldiers. Even during the starvation winters at Morristown and Valley Forge, Washington insisted that his quartermasters purchase provisions from the locals rather than resort to plunder and theft.[15] Even when Congress provided him martial law authority within thirty miles of his headquarters, Washington deferred legal matters to state and local governments if at all possible.[16]

Washington's success in maintaining and expanding loyalty to the Congress in the thirteen American colonies stands in stark contrast to the failed offensive into Canada, which Congress hoped to include in the new North American union. Though the Continental Army captured Montreal in 1775, the commander of the occupation force, an anti-Catholic bigot named David Wooster, instituted religious and requisition policies that so incensed the French Canadian majority that not even an emergency mission by Benjamin Franklin himself could salvage Canadian sentiments.[17] Unable to maintain a small American force in Canada in the face of local resistance, the Americans retreated south to their homes. Rather than gain an ally against England, Wooster had obtusely and remarkably turned the French Canadians into supporters of continued British rule, an outcome that not only endangered the rebellious colonies to the south in the Revolutionary War but also echoes across the further development of the American nation to this day.

The War of 1812

The War of 1812 was in many respects the end of the revolutionary era, a final chapter to determine what the geographic limits of American governance would be. As in the Revolutionary War, the American campaign to conquer Canada and include it in the United States was as halfhearted as the American resistance to British invasion from Canada was stout. The British participation in the war, ever on the periphery—at sea, along the Atlantic seaboard, in Florida, at New Orleans, in the Northwest Territories, in Canada—suggests the attacker once more measuring his quarry for a sign of weakness.

After defeating Napoleon in the Saxon campaign in 1813, England shifted forty thousand troops west across the Atlantic to prosecute a renewed offensive war in America. But in 1814, after the failed offensive from Canada, none other than the Duke of Wellington assessed British prospects for retaking America by force as slim. "I do not know where you could carry on . . . an operation which would be so injurious to the Americans as to force them to sue for peace."[18] The verdict of 1783 would stand: there was no American center of gravity, there would be no decisive battle, and entering a protracted war of attrition would ultimately mean victory for the Americans.

The Indians, always the primary focus of American military wrath, were again soundly defeated in major campaigns. The westerners who commanded the forces that fought the Indians into submission, William Harrison and Andrew Jackson, became the war's heroes and rode their fame to the White House.

The Mexican War

The Mexican War (1846–1848) was designed to be a strictly limited offensive war to define a new border between Mexico and the United States that favored American interests. The vast and remote western frontier areas of Texas south of the Neuces River and north of the Rio Grande, as well as New Mexico and California, were only sparsely populated with American settlers and adventurers intermixed with older and more established Hispanic outposts. At the time, California was home to a Hispanic population of only about eight thousand souls while the Anglo population probably numbered fewer than two thousand.[19]

The question at hand was whether Washington, D.C., or Mexico City would govern these border areas and their populations. President James Polk also entertained hopes that some of Mexico's other northern states would use the opportunity provided by the American military offensive south of the Rio Grande to secede from the Mexican republic and join the American union.[20] Attacking on six axes of advance, U.S. Army and U.S. Navy units secured all of the desired territories in the summer and fall of 1846, overwhelming the initial Mexican resistance, if any, that was offered.[21]

In California and New Mexico, the American guarantees of citizenship and retained property rights generally satisfied the populations in question, especially given their disaffection with the governance, or lack thereof, emanating from distant Mexico City.[22] Consequently, the military resistance offered by the annexed Hispanic populations was largely ceremonial and short lived.[23] The larger problem in the war was not conquering and pacifying the border populations but rather getting the Mexican government to recognize in a treaty the American fait accompli. Polk's dreams of securing Mexican states south of the Rio Grande were dashed by the poor civil-military conduct of Zachary Taylor's force advancing south from Texas.

The opening battles of the war occurred along the Rio Grande in Texas in territory disputed since the Texas War for Independence a decade before. In the battles of Palo Alto and Resaca de la Palma, the hastily assembled, largely volunteer American army led by Zachary Taylor clearly demonstrated its superiority in the field, inflicting consecutive lopsided defeats on the Mexican army. The stunned Mexican regiments retreated all the way to their operational base in Monterrey, where their American pursuers forced them to surrender.

President Polk expected the Mexican government to then concede defeat. Instead, General Santa Anna returned from exile in Cuba and rallied the Mexicans to counterattack in an attempt to expel the invading Americans. The poor conduct of Taylor's force—murder, rape, and looting were common—no doubt inspired the redoubled Mexican resistance.[24] The American side, escalating in turn, felt compelled to launch a second invasion force to fight its way into Mexico City and force the Mexican government to agree to peace terms. The accomplishments of Maj. Gen. Winfield Scott's small expeditionary army, invading outnumbered from the captured port of Veracruz into the Mexican heartland to seize the capital city, were testimony to American superior war-fighting abilities.

The Mexican government, ousted from its capital and with no real prospect of defeating the Americans, eventually agreed to cede the border areas in return for $15 million. Total American deaths from battle and disease exceeded thirteen thousand men, almost all of them in the prolonged campaign to Mexico City that the Americans thought they would never have to wage.[25] Mexican battle deaths are estimated at about five thousand.[26] Neither figure was demographically significant, the American population being around twenty million and the Mexican about seven million at the time.

For the nine months it took to negotiate and ratify the Treaty of Guadalupe Hidalgo, Scott's army had to govern Mexico City and its surroundings, as well as fight a war against Mexican guerilla bands contesting the American presence by attacking Scott's line of supply to the sea. Scott declared a war of extermination against the Mexican irregulars, executing many following summary trials. He ordered his troops to fine and eventually burn villages suspected of harboring guerillas. The road from Veracruz to Mexico City consequently became a "black swath of devastation several miles wide."[27]

While hard tactics honed in the Indian wars secured his lines of communication, Scott's occupation of Mexico City was a model of civility. Scott pledged to protect Mexican life and property, and he retained civic officials in their capacities wherever possible. He detailed five hundred American soldiers to serve as military police in concert with the Mexican police, thereby ensuring full and evenhanded enforcement of both Mexican and occupation laws.[28]

American troops distributed food to the poor, maintained public services and facilities, and paid the Mexicans fairly for requisitioned supplies. Scott ordered his troops to behave themselves properly and act with respect toward the citizenry, even having his soldiers salute Catholic clergymen.[29] The excellence of Scott's liberal military governance won over Mexican sentiments. Some Mexican liberals even proposed that Scott become Mexico's dictator as a step toward full assimilation of Mexico into the United States.[30] As Ulysses Grant, who served in the invasion and occupation, later remarked, "I question whether the great majority of the Mexican people did not regret our departure as much as they had regretted our coming."[31]

Scott was well prepared to impose effective and liberal martial law on his Mexican conquest. Trained in the law and well read, Scott realized that militarily appropriate rule of law in occupied territories would be as necessary as tactical victories if he were to achieve his objectives of capturing Mexico City and establishing a favorable peace. Mindful that the misconduct of Taylor's force had incited widespread resistance among the Mexicans, likely preventing the United States from annexing some of the north Mexican states, and also realizing that his small force would reach Mexico City only if it secured the cooperation of the Mexican population, Scott drafted his rules for the occupation, General Order 20, well before his force began its assault on the Mexican coast. Remarkably, General Order 20 applied equally to the Americans in his force and the Mexicans in his area of operations and was in essence a legal contract between occupier and occupied. Mexicans, individually and collectively, were expected to support the occupation, but they were also guaranteed personal and property rights.

Scott established military commissions, essentially summary courts, to decide on violations and punishments of occupation law.[32] Mexican and American alike learned early on that Scott was serious about the rule of law, even in wartime. Scott had an American civilian contractor convicted of rape publicly hanged in Veracruz in the opening days of the campaign.[33] During his advance to Mexico City, Scott's popular bilingual newspapers, established locally immediately after a city's capture, promulgated not only the rules of the martial law but also listed

convictions and punishments meted out to Mexicans and Americans for violations of occupation statutes.[34] Firm but fair rule of law, coupled with progressive
occupation policies, secured the flanks and rear of Scott's small army and allowed
it to concentrate its few combat regiments on the fighting.

In its scope and prosecution, the Mexican War was the only war America ever
fought that followed the Clausewitzian model: the limited objective, the border
battle, the destruction of the enemy army in a climactic battle of annihilation, the
seizing of the capital, the imposition of a peace treaty, and the military withdrawal.
In its own way, this war was a New World analog for the Franco-Prussian War,
which would be fought two decades later. (However, the manifest comparative
weakness of Mexico vis-à-vis the United States, unlike the peer status between
France and Germany, prevented Mexico from seeking later retaliatory wars.)

As Clausewitz warned, both sides found it difficult to rationally control the
war once the bloodshed began. Admirably, from a Clausewitzian perspective,
Polk maintained his original limited war aims rather than let his military success
carry him unthinkingly into ever-greater ambitions. Polk resisted overextending
America's reach and interest, annexing only the safest Mexican territory in the
north despite the growing sentiment in Washington and among some Mexican
elements for annexing more, some argued all, of Mexico.[35]

Three strategic lessons emerge from the war and occupation. The first is the
immediate need for both beneficent governance and rule of law in occupied territories. Both combatants and civilians must be governed by the rule of law as defined
in the published general orders of the commander. Citizens who cooperate with the
occupiers must enjoy the occupier's protection. Unduly harsh occupation, failure to
impose the necessary martial law, or failure to meet the basic economic and social
needs of the conquered population will only inspire resistance and bitterness. In
this respect, the civil-military conduct of Winfield Scott's and Zachary Taylor's
campaigns stands in distinct contrast in both conduct and outcome.

The second lesson is the need for unrelenting counterinsurgency operations in the wake of conventional operations. Again, counterinsurgency
operations must be governed by and subordinate to the commander's imposed
rule of law as defined in the occupation statutes. Insurgents and rebellious populations lose legal protections only as required by military necessity and only as a
consequence of their failure to comply with occupation policies.

The final lesson is that the same commander must be responsible for all three
interrelated aspects of the war: the conventional campaign, the counterinsurgency,
and military governance. Ultimately Scott understood that he could not tell the
Mexicans anything different from what he was telling his troops and, consequently, the campaign had to be waged with a single voice—that of the overall
commander. The message, moreover, had to be consistent in time and space.
General Order 20 would apply throughout his area of operations for all people

from the moment the Americans came ashore until the time they debarked a year later to return to America.

The precedent of command unity set by Scott in Mexico would echo through the American military pattern for the next one hundred years and lead to historic successes on an ever-widening world stage. Our disregard of this precedent from Vietnam forward has led only to frustration and defeat.

The American Civil War

The American Civil War was a classic war of attrition in which the North, on the offensive from the beginning, sought to impose unwanted governance on the South. The overwhelming sentiment of the people and public officials in the Southern states after the election of Abraham Lincoln in 1860 was for secession. Lincoln and the North were determined to defeat the Confederacy and bring its states back into the Union. It was not the South's intention to change the government in the North but rather to defend its new Confederacy.

Militarily, neither side at first envisioned the total war that would develop. There was hope that an initial large battle would settle the matter favorably, but the Union defeat at Bull Run led only to fuller mobilization on both sides and promises of even bigger battles and campaigns in the future. Both sides were convinced that the center of gravity in the war was the corridor connecting the two capitals, Washington and Richmond, and the two armies opposed there, the Army of the Potomac and the Army of Northern Virginia. But despite the best efforts of the generals on both sides, there would never be a decisive battle in the east but rather an unrelenting series of bloodlettings that the North could withstand longer than the South could.

The Civil War illustrated the primacy of demographic factors in war and served as the benchmark for American war doctrine for the next hundred years. The opening battles, informed by Napoleonic strategies, achieved no consequential results and merely fueled the war fever, leading to total mobilization of the military manpower on both sides. Every summer in the Washington-Richmond corridor, in the Tennessee and Mississippi river valleys, and along the Gulf and southern Atlantic coasts, the opposing armies would fight each other into exhaustion in a seemingly endless series of battles, many now lost to all but the most ardent historians. As the Union Army slowly advanced, bitter guerilla warfare in the border states, especially Missouri, forced the Federals to conduct simultaneous conventional and pacification offensives. Over time, the North adopted the hard-war strategy, best summarized by quotations from one of its chief architects, Gen. William Tecumseh Sherman:

> War is the remedy our enemies have chosen, and I say let us give them
> all they want . . . We are not fighting armies but a hostile people, and

must make young and old, rich and poor, feel the hard hand of war . . . I would make this war as severe as possible, and show no symptoms of tiring till the South begs for mercy . . . My aim, then, was to whip the rebels, to humble their pride, to follow them to their inmost recesses, and make them fear and dread us. Fear is the beginning of wisdom . . . Every attempt to make war easy and safe will result in humiliation and defeat.[36]

In practice, this meant not only destroying Southern manpower in battle but also destroying the economic base of the Southern war effort and making the war intolerable to its civilian supporters. Sherman's infamous sacking of Atlanta and subsequent March to the Sea are only the best-remembered highlight of the hard-war policy.

Firm punishment of Southern insurgents behind Union lines was the standard practice. As part of the Shenandoah campaign, Gen. Ulysses Grant ordered the arrest of all families of known guerillas as well as "all able-bodied male citizens under the age of fifty . . . suspected of aiding, assisting, or belonging to guerrilla bands." Grant also considered forced removal of the entire population of northern Virginia because of "the necessity of cleaning out that country so that it would not support Mosby's gang" of irregulars.[37] Later in the same campaign, Gen. Philip Sheridan wrote in his October 7, 1864, report to General Grant:

I have destroyed over 2,000 barns filled with wheat, hay and farming implements; over 70 mills filled with flour and wheat, and have driven in front of the Army over 4,000 head of stock and have killed and issued to the troops not less than 3,000 sheep . . . people here are getting sick of the War, heretofore they have had no reason to complain because they had been living in great abundance . . . Tomorrow I will continue the destruction down to Fisher's Mill. When this is completed, the Valley from Winchester to Staunton, 92 miles, will have but little in it for man or beast . . .[38]

Killing or removing hostile people, and capturing or destroying the materiel that sustains them, is what war is all about. From the Union Army's standpoint, the only question was how much they were going to have to hurt the South before it abandoned its brief independence and long-held institution of slavery. The South reached its breaking point in the spring of 1865.

For over four years the two sides had fought countless battles and lesser engagements in every Southern and border state as well as some in the North. Much of the South was in ruins; in places its population was on the verge of starvation. On the Confederate side 1 million men were mobilized from a white population of 5.6 million, almost 18 percent, one of the most complete mobilizations ever recorded. Over the course of four years of battles, skirmishes, and hardships, 258,000 Southern soldiers died, 189,000 were wounded, and 215,000 were taken prisoner. Even considering double counting—that is, soldiers who fall into more than one category—the Confederate military casualty rate was well over 50 percent, or 9 percent of the overall Southern white population. No reliable statistics of Southern civilian deaths are available, but published estimates of 50,000 Confederate and border-state citizens, something less than 1 percent of the population, do not seem unreasonable given the Northern hard-war strategy and widespread malnutrition. The combined Confederate death rate, both military and civilian, was over 5 percent.

Northern military casualties in all categories were almost 650,000, about one-tenth of the adult male population, or 3 percent of the overall population of 22 million. The death rate was about 2 percent, similar to that of the Revolution.

The North mobilized 2.1 million men for the Civil War, slightly less than 10 percent of its population. Eighty percent of the Northern mobilization was males between ages eighteen and twenty-nine; middle-age men and boys were rare in the ranks. One in ten Northern able-bodied adult males was killed or wounded in the war.[39] The comparatively lower rates of Northern mobilization contributed to the great strength of the Union war economy, and the lower casualty rate promoted the willingness of the electorate to see the war of attrition through to the end. Northern civilian casualties were negligible.

General Robert E. Lee's letter to his troops, written the day he surrendered the remnants of his once-proud army, bespeaks the inability of Southern manpower to continue the war.

> After four years of arduous service marked by unsurpassed courage and fortitude, the Army of Northern Virginia has been compelled to yield to overwhelming numbers and resources. I need not tell the brave survivors of so many hard fought battles, who have remained steadfast to the last, that I have consented to this result from no distrust of them; but feeling that valor and devotion could accomplish nothing that could compensate for the loss that must have attended the continuance of the contest, I determined to avoid the useless sacrifice of those whose past services have endeared them to their countrymen.[40]

The success of the Union Army in exhausting the South's military-age manpower and economic means forced the Confederate surrender. The South was beaten and she knew it. Peace with the North was the only option.

The North's occupation and military governance of the South had already begun prior to Lee's surrender. As the Northern armies fought their way south, considerable Southern populations came under Union control, and groups of Confederate sympathizers often pursued their cause through guerilla warfare. General Orders 100, *Instructions for the Government of Armies in the Field*, signed by President Lincoln himself, provided a legal code for the conduct of the army toward enemy regular forces, irregular forces, and peaceable civilians, along with a combination of the laws of land warfare, counterinsurgency policy, and military governance rolled into one.

General Orders 100, also known as the Lieber Code after its chief architect, was the first significant attempt in either Europe or America to clearly define the "laws of war," and it quickly received international acclaim.[41] It served as the basis for the Hague Conventions of 1899 and 1907 and the Geneva Conventions of 1947.[42] In very brief summary, battle between uniformed armies would be similar to European norms, guerilla fighters could be summarily executed, populations that supported guerillas could be punished with the severity necessary to stop their support, and law-abiding civilians would enjoy most basic rights (property, commerce, justice, religion) as long as they did not oppose the military government.[43] A progressive document, General Orders 100 frames warfare in terms of enforcing a just conclusion to hostilities rather than exacting retribution through military force. "[I]t is incumbent upon those who administer [martial law] to be strictly guided by the principles of justice, honor, and humanity—virtues adorning a soldier even more than other men."[44]

The generally fair governance of the occupied South, given the circumstances, and Lincoln's eloquent assurances of a just peace throughout the South following surrender were indispensable elements of the North's overall war strategy.

The Union Army's military governors, usually serving Union officers appointed by Lincoln himself, exercised a wide range of executive, legislative, and judicial powers in occupied areas. They supervised, removed, and appointed local governmental and civic leaders as necessary. They supervised elections and controlled the press. Following Scott's example from the Mexican War, they created military commissions and martial courts to punish offenders of occupation law and supervised existing civilian courts trying nonmilitary offenses. By so doing, the same military commanders who tried Union soldiers at courts-martial for offenses against Southern civilians also tried civilians for offenses committed against the Union Army and its cause.

Military governors also exercised authority over economic matters and public finance within their occupation districts. Indeed, every sphere of public activity, including ecclesiastical affairs, could come under the purview of the military governors as required by common sense and military necessity.[45]

The army occupied the defeated South from 1865 to 1877. Initially, the army operated military governments under martial-law authority throughout the South, often using local officials to administer the various functions of government.

Security and restoration of normalcy, except in the case of the freedmen, was the goal. The democratically elected Southern state governments were initially recognized by the Andrew Johnson administration, though their powers were ill defined and limited by the local occupying military authorities.[46]

After the 1866 elections, however, the Radical Republicans reversed state recognition and restored enhanced military rule, seeking the removal of former Confederate leaders, the disestablishment of local militias, and much deeper political change in the South.[47] The Congress administered the South through five newly created military districts, albeit with decreasing army forces. The two hundred thousand soldiers on occupation duty in the summer of 1865 were reduced to twenty thousand by the fall of 1867 and nine thousand by 1870, when the last of the former Confederate states was readmitted to the Union.[48] Several thousand U.S. troops remained in the South to enforce Federal policies until 1877, when political reconciliation finally brought Reconstruction to an end. The costs of maintaining a large occupation force, plus the realization that only so much social transformation, especially regarding racial attitudes, could be accomplished by military government, made continued occupation impractical. As General Sherman stated, "No matter what changes we may desire in the feelings and thoughts of the people [in the] South, we cannot accomplish it by force."[49]

However incomplete in building racial harmony, the Reconstruction, from a military strategic perspective, can only be viewed as an unqualified success. Despite the hatreds and hardships of the war, in the course of a decade the majority of Southerners transferred their allegiance, if not their sympathies, from the Confederacy to the remade United States and got on with building the common future, almost as if the war had never happened. The South collectively found it remarkably easy to reassume its identity as part of the American union, even if it could not accept the new racial equality that the war in some measure tried, but failed, to achieve.

The Spanish-American War

The American desire to support the Cuban insurrection against Spanish colonial rule directly led to war in 1898. Upon Congress' declaration of war, the U.S. Navy, with its far superior fleet, sailed against the decrepit Spanish naval squadrons in the Philippines and Cuba and sank or captured all sixteen Spanish warships in the two island colonies. Only one U.S. sailor was killed and nine were wounded in the exercise. The Spanish had no means to reverse the decisions in Manila Bay and Santiago Harbor; its remaining navy was no better than what had already been lost.[50] Unable to maintain its island colonies without sea lines of communication, the Spanish monarch had no option but to cede them in the Treaty of Paris. Our war with Spain was textbook limited war.

The army's liberation (then occupation) of Cuba is of more interest to our study. Cuba had been devastated by nearly four years of insurrection and war when the U.S. Army invaded Cuba in support of the Cuban *insurrectos* in the summer of 1898. United States Army units and *insurrectos* fought side by side in the brief but decisive battles that defeated the Spanish forces and led to the Spanish surrender.

Despite the desires of the Cubans for immediate self-rule, the Americans judged that the Cubans had no experience with self-governance and saw the need for the U.S. Army to maintain order during the transition from Spanish colonial rule.[51] Accordingly, the Americans negotiated an agreement with Spain to take over governmental control in Cuba from the departing Spanish in January 1999. With only eleven thousand American soldiers assigned to Cuban occupation duty, imposing order on 1.5 million Cubans, now anxious to enjoy their independence, was no easy task. The goodwill the Cubans felt for their liberators in some measure aided the occupation. But the tremendous demographic losses that Cuba had suffered in its war against Spain, three hundred thousand dead (16 percent), and the accompanying collapse of the Cuban economy and financial system had left Cuba no option but to accommodate its rich and powerful occupiers from the north.[52]

The army commander, Maj. Gen. John Brooke, established military government using local governmental agencies and their staffs, insofar as possible, with Americans in decision-making roles. Brooke also got the fifty-thousand-man Cuban Army of Liberation to agree to demobilize by paying their wages, buying their weapons, and hiring thousands into the military government or the new American-controlled Rural Guard. Co-opting the Cubans, rather than fighting them, proved to be the key to establishing law and order.

Brooke also began delivering the usual assortment of public works programs, including sewers, roads, and schools, that were typical of American pacification campaigns. These efforts were further accelerated by Brooke's successor, Maj. Gen. Leonard Wood, who also instituted political and legal reforms designed to prepare the Cuban people for self-rule. In May 1902, after three and a half years of military government, the U.S. Army finally handed over all authority to the Cuban Republic and sailed home.[53]

But our first victory in overseas nation-building was short lived. In 1906, six thousand American troops were ordered back to Cuba when a disputed election led to a new revolution. American soldiers and marines occupied the island without opposition and began three more years of military government, civic works projects, and progressive reform measures. Still, after the Americans left in 1909, Cuba could not maintain its democratic institutions and eventually descended back into corruption and dictatorship.[54] Despite the U.S. Army's best efforts, Cuba would not achieve democratic self-rule.

The Philippine War

The army achieved more lasting success in the Philippine occupation, despite, or perhaps because of, the bloody insurrection and pacification that followed Spain's leaving. In contrast to the largely peaceful and mutually cooperative liberation of Cuba, the Filipinos militarily resisted our occupation of the Philippines and our declaration of sovereignty over the archipelago. The Filipinos, much like the Cubans, had fought a guerilla war against Spanish rule throughout much of the 1890s. They expected their reward to be independence once the Spanish withdrew. However, America's decision to keep the Philippines as a U.S. territory caused the Filipinos, led by Emilio Aguinaldo, to declare their independence from America and organize for defense.

Americans were at first unaware of how unpopular our conquest would be or of the depth of commitment among the Filipino people for their independence. As Brig. Gen. Arthur MacArthur, who later would become the American military governor, told a visiting reporter in the fall of 1899:

> When I first started in against these rebels, I believed that Aguinaldo's troops represented only a faction. I did not like to believe that the whole population of Luzon—the native population that is—was opposed to us and our offers of aid and good government. But after having come this far, after having occupied several towns and cities in succession, and having been brought much into contact with both insurrectos and amigos, I have been reluctantly compelled to believe that the Filipino masses are loyal to Aguinaldo and the government which he heads.[55]

The war opened with a brief insurrection against the American occupation in Manila, followed by a yearlong series of conventional battles against Aguinaldo's army, and ended with a prolonged guerilla war. As in our other wars, occupying centers of gravity and winning battles of annihilation failed to produce a quick decision, and attrition war became the only path to success. While Manila and other areas succumbed to policies of "attraction," "chastisement" became the default strategy in rebel areas. Only the army's reversion to the hard-war tactics honed during the Civil War and Indian campaigns would pacify insurgent Philippine provinces. Villages were burned and hundreds of thousands of rebel supporters were temporarily relocated into "protected zones," similar to Indian reservations in America. In one area of Mindanao, all military-age males were removed from the population.[56]

The "masterpiece" of the Philippine counterinsurgency was a five-month campaign waged by Brig. Gen. Franklin Bell's 3rd Brigade in the Luzon island provinces of Batangas, Laguna, and Mindoro from December 1901 to May 1902. Bell, who had studied law and was a member of the Illinois bar, cited Lincoln's General Orders 100 as his authorization to employ whatever reasonable means necessary to bring the *insurrecto*s in this untamed region into a state of obedience to their new sovereign, the U.S. government.[57] Bell provided a copy of the General Orders 100 to every officer in his brigade so that all would be impressed by the legal basis of their actions. He published his own general orders in written form, and ensured that the Filipinos in his area of responsibility understood the rules of the occupation and the price of breaking them. Combining martial law with tough soldiering and thorough population-control measures, Bell brought the insurrection to an end with surprising quickness.

Bell's techniques were both enlightened and severe. He controlled all food storage and sales in the provinces, outlawing and destroying all food stocks in the countryside. This forced the peasants into the cities, which were controlled by American garrisons, where they could be isolated from the insurgent bands they supported. Food in the countryside could be purchased only in daily rations from American-approved commissaries, which sold foodstuffs taken from wealthy insurgent supporters.

Bell kept much of his force aggressively hunting the enemy in the hinterlands, inflicting casualties on them and forcing them to keep on the move, denying them sanctuary and wearing them out physically. Military-age males who ran from American patrols were shot.

Filipino civil authorities and wealthy citizens were forced to overtly and publicly support their American occupiers; those who refused were arrested. Bell understood that the Filipino elites, not the common hombres, were the leaders and funders of the insurrection, so co-opting their support became the focus of his efforts. His message to the elites was clear: support us or lose your privileged positions.

Filipinos who attacked Americans or *Americanista*s (Filipinos who supported the Americans) were executed after summary court hearings. Should the guilty party not be found, imprisoned insurgent leaders would be executed in their place. Bell personally made all death penalty decisions. Towns that resisted American occupation would be "completely destroyed by fire as a measure of retaliation."

Bell's methods brought immediate results. Within five months his forces had rolled up the entire *insurrecto* force and captured its leaders. Of the 4,000 insurgents, only 350 were killed or wounded and another 900 were captured. Nearly 3,000 surrendered under conditions granting amnesty and parole in return for their allegiance. However, the countryside was "devastated." Moreover, malnutrition and disease, a direct consequence of Bell's rationing and concentration methods, caused significant hardship and death among the general population, forcing Bell eventually to relax his food restrictions and implement public health measures and a program of inoculations against communicable diseases.[58]

At the peak of the insurrection in 1901, some seventy thousand U.S. soldiers were deployed in the Philippines. Over fifteen thousand Filipino auxiliaries, organized by the Americans into scout and constabulary units, augmented the

American force.[59] Perhaps twenty thousand of the one hundred thousand Filipinos who took up arms against the United States were killed in the war. Another fifteen thousand noncombatants died from military action. Nonbattle casualties, especially from disease, were far higher. Although some sources believe that Filipino deaths in all categories may have been up to a million, the most common estimate of Filipino civilian casualties is two hundred thousand out of a population of about seven million.[60] Even the lower figure indicates an overall death rate of about 3 percent. Of course, death, injury, economic hardship, and incarceration were disproportionately felt in rebel areas, where casualties in some cases approached the same intolerable levels reached in America during the later Indian Wars.

American military government was the umbrella under which the counter-insurgency operations took place. For the first two years, the army directly ruled the Philippines using martial law. As in Cuba, America engaged in civic action programs wherever the military situation allowed. Military government meant schools, roads, legal systems, and democratic reforms. Beginning in late 1900, pacified areas were removed from direct military control and fell under President Theodore Roosevelt's appointed civilian governor, William Taft. Taft issued 499 statutes that reversed Spanish law and formed the basis for the new territorial law. He also made English the official Philippine language.

In December 1899 the Americans allowed the formation of the pro-American Federal Party, headquartered in Manila, as a Filipino alternative to Aguinaldo and the insurrectionists; for a variety of reasons, ranging from the idealistic to the self-serving, 150,000 Filipinos joined the party in its first few months. As the political nucleus of a new Filipino identity, the Federal Party was instrumental in providing local civilian government officials in pacified areas where the Americans allowed it, and negotiating the surrender and co-opting of insurgent leaders.[61]

After Roosevelt declared the war over in 1902, Taft allowed local and provincial elections; however, Americans retained all executive authority. In 1907 the Filipinos voted for their first national legislature, though the upper house and executive branch still remained American presidential appointees. Only 3 percent of the adult population voted. The overwhelming winner was the Nationalist Party, which demanded "complete, absolute and immediate independence."[62]

The 1916 Philippine Autonomy Act announced American intent to eventually grant independence to the Philippines rather than retain it as a territory for eventual statehood. The new Philippine constitution provided for Filipino election of both houses of the legislature, but retained the executive branch for the Americans and their appointees.

In 1934 the new Franklin Roosevelt administration finally set the date for Philippine independence: 1946. In 1935 the Filipinos elected Manuel Quezon their first president. However, an American, Gen. Douglas MacArthur, was appointed by Roosevelt to command the Philippine army.

Finally, in 1946 the Philippines were granted their independence, although their military stayed linked to ours through alliance and basing agreements. Only after the end of the Cold War did we give up our Philippine bases. This litany of events is remarkable in illuminating the step-by-step process through which an occupier gives up his instruments of power while trying to remain reasonably assured that his interests will be met by the new government.

The Philippine post-independence democracy has not been completely steady, suffering periods of crisis and a decade and a half of military rule under the regime of President Ferdinand Marcos. Nor did the American occupation put the Philippines on the track to prosperity; it remains one of the poorest countries in East Asia. Still, the Philippine campaign can only be considered a triumph of offensive warfare and counterinsurgency—about as good as it gets—and deserves more attention than the army currently pays it.

World War I

The madness that caused World War I is too complicated and nuanced to describe in a brief summation. All the nations involved believed that they were defending against the aggressive designs of their enemies, yet all had strategic aims that could be realized only by victory in war. All had prepared alliances, mobilization plans, and military campaigns that they believed would lead to their victory but, collectively viewed, would lead only to their mutual defeat. All anticipated that the war would be won in decisive opening battles of annihilation. Had they accurately forecast the attrition war that was to consume them, all certainly would have opted for other ways to resolve their differences. The war demonstrated the inappropriateness of monarchical rule in the new democratic age, and all the major dynasties of Europe were swept away in the revolutions that accompanied the war.

Despite Germany's Clausewitzian opening strategy, the country could not repeat its victory of 1870 over France, and World War I quickly became a war of attrition on two fronts. Over the four years of the war, Germany mobilized 11 million men out of a population of 65 million (17 percent), the virtual demographic limit.[63] In countless, many nameless, engagements, battles, and campaigns, 2 million German soldiers died, 4.25 million were wounded, and 1 million were captured. Though the war rarely came to German lands, civilian war deaths, mainly from malnutrition, numbered 762,000, over 1 percent. Fully 10 percent of the German population was killed, wounded, or captured in the war. This was enough bloodshed to cause German military and political collapse and force an armistice, but, as we all know, it was not enough to compel a permanent peace.

France suffered even more. She mobilized 8.4 million soldiers, of whom 1.4 million died, 4.3 million were wounded, and a half million were made prisoners of war. Fully 11 percent of France's prewar population was either killed or wounded by war's end; most all were military-age males. Extreme casualties brought the

French army to the point of mutiny in 1917. France had simply run out of new men to throw into the fight and leaned increasingly on the British and Americans to man the fronts she could no longer defend.

Russia mobilized more men than any other combatant, 12 million from a population of 174 million (about 7 percent); 1.7 million soldiers died, 5 million were wounded, and 3.6 million were made prisoners of war. Two million civilians died as well. The incredible demographic and economic toll of the war ultimately collapsed the tsarist regime and forced Russia to abandon its war effort as the country descended into anarchy and revolution. Russia may have fractured first, but there was a foreboding throughout all of Europe's empires—Austro-Hungarian, Ottoman, and German—that they were engaging in the same insanity toward the same end. And indeed they were.

The American entry into the war was the straw that broke the back of the Triple Entente. Declaring war in 1917, the Americans mobilized only about 4 million soldiers for World War I, a fraction of our demographic potential, and only about half made it over to France by November 11, 1918, Armistice Day. But the effect was immediate; the first arriving divisions blunted the Germans' do-or-die 1918 spring Ludendorff Offensive. By the Fourth of July, 1918, the American Expeditionary Force in France numbered a million men, and 250,000 additional U.S. troops were arriving in France each month. These troops, in combination with the British in Flanders, would launch the fall counteroffensives that would crack the weakly held German lines. Riots erupted in Germany, the navy mutinied, and soldiers began organizing revolutionary councils, a replay of Russia the year prior. Ludendorff, now the German chancellor, fled to Sweden, and the kaiser abdicated and fled to Holland, both unwilling to be held accountable for the calamities they had caused.

There was no general occupation of Germany after its collapse ended the war. France took back the provinces it lost in 1870. The allies, including the Americans, briefly occupied the German Rhineland to enforce the terms of the armistice, but there was no real attempt to impose a new form of governance on any of the defeated belligerents through military occupation. The only nation with the remaining energy to do so, the United States, was at this point unwilling to play this role on the world stage. The war, as it turned out, was a limited war fought with unlimited means, a tragic war whose ends did not justify its means. For this book it is instructive only because it provides evidence concerning achievable levels of mobilization that modern nations can attain and the percent of casualties they can tolerate before they break.

World War II

World War II began with Axis offensives aimed at building new empires through permanent occupation, the Allies initially fighting defensive wars of preservation.

By 1943 the Allies had recovered from their early defeats and were on the counteroffensive, successfully liberating previously occupied territories and countries. The 1943 Allied declaration of the unconditional surrender policy dramatically changed the war, forcing the Axis powers to defend their regimes against ever more powerful Allied offensives whose stated purpose was nothing less than the military occupation and political reconstruction of both Germany and Japan. Rarely in history have the tables been so completely turned. Instead of fighting for empire, the Axis governments ended up fighting for their very survival. The conqueror ultimately became the conquered.

For Germany, World War II was a national effort to raise, equip, and employ a new generation of soldiers to avenge its World War I defeat. Unlike the first war, Germany's goal was total, nothing less than the conquest of Europe and the installation of Nazi governance in all the occupied lands. Unlike 1914, the decisive opening campaigns succeeded in defeating Poland and France with relatively light German casualties. To Hitler's surprise, the follow-up attack to the British Isles failed. Nevertheless, Germany invaded the Soviet Union in 1941, and again enjoyed spectacular initial victories. The hugely successful Soviet counteroffensive at Moscow, the first costly German defeat on the ground, and the American entry in the war, both in December 1941, forced a war of attrition that would bring Germany to its knees over the next three and a half years. Trapped again in a two-front war against its more populous and powerful enemies, the German nation lived its worst strategic nightmare.

In World War II Germany lost 3.25 million military dead, 7.25 million military wounded, and, at war's end, had 7 million soldiers in POW camps, accounting for almost of all of its military-age male population, as well as 1.2 million civilian deaths.[64] Death rates approached 7 percent of the population, and combined casualty rates approached 20 percent, rates about 65 percent higher than those in World War I. Moreover, the Allied bombing campaign and ground combat in Germany caused widespread economic destruction, with 39 percent of the dwellings in Germany's forty-nine largest cities destroyed or severely damaged by war's end. Starvation and destitution gripped the land. Germany's World War II surrender was unconditional and permanent, reflecting the severity of its casualties and its economic calamity. There would be no popular opposition to the American occupation.

Admiral Karl Doenitz, the president of the German Reich at the time of the surrender, reflected the hopelessness of the German situation in his last radio address to the German people:

> German men and women: When I addressed the German nation on May 1 telling it that the Fuehrer had appointed me his successor, I said that my foremost task was to save the lives of the German people. In order

to achieve this goal, I ordered the German High Command during the night of May 6-7 to sign the unconditional surrender for all fronts.

On May 8 at 23 hours (11 P. M.) the arms will be silent.

German soldiers, veterans of countless battles, are now treading the bitter path of captivity, and thereby making the last sacrifice for the life of our women and children, and for the future of our nation.

We bow to all who have fallen. I have pledged myself to the German people that in the coming times of want I will help courageous women and children, as far as I humanly can, to alleviate their conditions. Whether this will be possible I do not know.

We must face facts squarely. The unity of state and party does not exist any more. The party has left the scene of its activities.

With the occupation of Germany, the power has been transferred to the occupying authorities . . .

All of us have to face a difficult path. We have to walk it with dignity, courage and discipline which those demand of us who sacrificed their all for us. We must walk it by making the greatest efforts to create a firm basis for our future lives.

We will walk it unitedly. Without this unity we shall not be able to overcome the misery of the times to come. We will walk it in the hope that one day our children may lead a free and secure existence in a peaceful Europe . . .[65]

The Allies rejected Doenitz' offer to "help" and imprisoned him for war crimes. The Allies were determined to rule and remake Germany without the assistance of the country's former Nazi leaders.

In March 1945 Gen. Lucius Clay was made Eisenhower's deputy commander for military government, with specific responsibilities for governing the twenty million Germans in the American zone. Importantly, from October 1945 on, Clay ensured that military governance would fall under its own command, the Office of Military Government, and not be the responsibility of tactical unit commanders and their G-5 (civil affairs) staffs.[66] Separating military governance from tactical command was strongly recommended in army doctrine based on lessons learned from the post-World War I Rhineland occupation.[67] The core of the occupation effort was the Office of Military Government staff, which numbered 12,000 soldiers and civilians in December 1945 and was gradually reduced to 2,500 by January 1949 as American confidence in the new German government improved.[68]

The army's military governance organization did not emerge spontaneously in the spring of 1945 but was instead the result of years of deliberate planning. In 1940, over a year before the attack on Pearl Harbor, the army anticipated the eventual need to defeat and occupy the Axis powers and issued Field Manual 27-5, "Military Government," which was largely based on lessons learned by the Progressive Era army in the occupations following the Spanish-American War and in the Rhineland after World War I.

On April 2, 1942, during the post-Pearl Harbor buildup, the secretary of war established the School of Military Government at the University of Virginia at Charlottesville. Its charter was to refine military government doctrine and train the many thousands of civil affairs officers and enlisted men who would be needed once the liberations and occupations began in Europe and the Pacific.[69] For the first time in its history, the U.S. Army had created a dedicated and specifically trained military governance and civil affairs corps to administer civilian populations in the wake of combat operations.

In 1942, a few months into the defensive phase of the war, the army had created a special group in England, consisting of hundreds of commissioned officers and civilian experts, to examine all aspects of German governance and prepare the groundwork for the occupation.[70] By D-Day over eight thousand American civil affairs soldiers were staged in England; they were already trained and organized to perform military governance as the occupation of Germany unfolded.[71]

The basic concept of the occupation of Germany was to unroll a carpet of military governance behind the advancing "fringe," to use Eisenhower's term, of combat troops. Military governance detachments would advance with the combat units headed toward their predetermined jurisdictions of operation, but upon arrival they would detach themselves from the combat forces and stay behind to serve as military governors. In short order the detachments' command chain would run not through the combat divisions and corps but through a special command established for governance of the rear areas and then, after the hostilities ceased, through the deputy commander for military government, General Clay.

The army's tactical battle drills for occupying a village or town were well trained and rehearsed:

> The procedure was the same everywhere, as it was to be throughout Germany. First came the posting of the Supreme Commander's proclamation [of occupation and martial law] and the ordinances [of the military government]. The second step was to find the Buergermeister (mayor) or, if he could not be found or was obviously a Nazi, appoint one and thereby establish a link to the population. Next came a series of security actions. The first was to collect weapons, ammunition, and explosives in civilian possession and confiscate radio transmitters and other means of communicating with the enemy, including pigeons. The orders to surrender prohibited items were followed by house-to-house searches,

which in fought-over areas frequently turned up sizable collections of arms that the civilians had not turned in, probably more out of fear than malice. For convenience and for security, the civilians also had to be kept out of the way of the tactical troops.[72]

The army had already determined, translated, and mass-produced its initial Proclamation and Ordinances before the occupation of Germany even began. In accordance with FM 27-5, civil affairs detachments posted these documents immediately behind the advancing line of contact; indeed, word of their content often preceded the arrival of American troops.

The proclamation, ordinances, and laws—also printed separately in large format for posting—would constitute the legal bond between the Germans and military government. Although not strictly required in international law, the proclamation was assumed to be accepted United States practice. Addressed to the people of Germany in the name of General Eisenhower as Supreme Commander, Allied Expeditionary Forces, it declared his assumption of "supreme legislative, judicial, and executive power within the occupied territory"; suspended German courts and educational institutions; and required all officials and public employees to remain at their posts until further notice. The first of the three ordinances defined nineteen crimes against the Allied forces punishable by death. The second ordinance established military government courts, and the third made English the official language of military government. The laws, with gaps left in the numbering system to accommodate future legislation, fell into two classes: those necessary to establish and maintain military government control and those dealing with national socialism. Law No. 1 abrogated nine fundamental Nazi laws together with their subsidiary decrees and regulations and prohibited any interpretation of German law in accordance with Nazi doctrine. Other laws abolished the National Socialist Party, its auxiliary organization, and the use of its emblems.[73]

As Germany collapsed in the spring of 1945, thousands of combat and support soldiers were diverted to military government tasks, so pressing was the need to deal with displaced persons, the destroyed infrastructure, and the effects of governmental and economic collapse. One corps alone reported in April 1945 that it had diverted about six thousand soldiers to military government duties, "dissipating" its combat effectiveness.[74] After combat operations ended in 1945, several U.S. tactical units were reorganized and retrained as a thirty-thousand-man constabulary force for police and riot-control duty, with the planning factor being 2.2 constabulary troopers per thousand population.[75] Only a small number of soldiers assigned to combat divisions ever engaged in military governance activities after 1945, their mission being either to deploy out of Germany or to stay prepared to fight conventional war, increasingly with the Soviets in mind.

Germany was politically broken asunder at war's end, with no functioning national government and no real Allied agreement on forming one. Moreover, the political class that had ruled Germany in recent memory had either perished in Nazi purges or was now complicit, perhaps criminally, in Hitler's regime. Even though purging Nazis from all levels of German governance was a priority, Clay realized that governance in the American zone alone required three hundred thousand government workers, almost all of them of necessity German citizens, so he had to quickly develop mechanisms to coordinate governmental activities with whatever local officials could be trusted, even former Nazis.[76]

Unable to impose governance from the top down because there was no German national government, Clay had to work from the bottom up. In July 1945, two months after the surrender, army military government teams in the field restored German city and county governments in the American zone, placing German officials under strict American supervision. In September 1945 the state governments were reestablished under German administrations headed by American-appointed minister-presidents, again closely supervised by American liaison teams.[77]

The Americans immediately assumed responsibility for maintaining order and imposing rule of law. At the time of initial occupation, the Americans disarmed (but did not disband) the local police forces, allowing them to patrol using nightsticks under American supervision. By September 1945 the remaining German police, thoroughly de-Nazified, were rearmed with American carbines, and by October there were twenty-two thousand German police working closely with American military police maintaining order.[78]

Also, at the time of occupation, German courts were disestablished and Americans imposed martial law, meaning that German civilians would be tried in U.S. military courts. By June 1945 nearly five hundred U.S. military government courts had been established to try offenses committed against occupation authorities.[79]

Trials using German juries were not possible at this time, so American officers conducted trials by summary courts open to the public; American officers decided the cases. Clay believed that the open military governance courts conducted using American judicial standards served as valuable demonstrations of American commitment to rule of law and justice, and effectively modeled the democratic justice system to the German population. By August 1945 de-Nazified German local courts were reestablished for minor offenses, though their decisions were reviewed by American lawyers, and in 1946 German jurists working with the occupation authorities completed a post-Nazi era administrative law. In 1947 state constitutions were adopted ensuring Germans equality before the law, due process, speedy and public trial, and the right to confront witnesses.[80] In August 1948 the U.S. military government finally handed all jurisdiction over German citizens back to the reformed German courts. In the nearly four years since American soldiers first crossed the German border, U.S. Army military courts had tried almost four hundred thousand cases against German defendants.[81]

Over time, German governmental capacity grew, and the Americans became convinced that the Germans could rule themselves democratically. As was the American historical practice, sovereignty was restored to Germany in stages, each linked to the success of the previous test. The first free elections in Germany since 1933, the last year of the Weimar Republic, were the American zone village elections in January 1946, followed by the county and small-city elections in April and the large-city elections in May. Participation was 70 to 80 percent, with a majority of the American appointees returned to office. Importantly, in the German popular assessment, Clay hadn't forced appointed officials into the role of collaborators but had allowed them to represent the interests of the citizenry. State constitutional assemblies were elected in June, and the new constitutions were submitted to the military government for approval in October. By year-end 1946, the electorates of all three American-zone states had ratified their constitutions and elected their legislative representatives and governors.[82]

Although Clay, as the occupier, could overrule any German governmental action, the manifest cooperation between the Americans and Germans that emerged during the postwar period made such action unnecessary. The 1949 formation of a West German federal government out of the states of the American, British, and French occupation zones restored German nationhood, albeit with occupier rights and extraterritoriality. Not until 1955 was Germany allowed to reestablish its own military, and then with the strict understanding that it would be employed only under North Atlantic Treaty Organization (NATO) command exercised by its Allied occupiers.

American generosity was the single greatest factor in our occupation success in Germany. When Germans were faced with starvation, the Americans responded with food, even if guaranteeing only a 1,550-calorie per day ration. The American school lunch program fed the German children in the American and British zones through the starvation winter of 1946–1947. In Clay's words, "The child feeding program did more to convince the German people of the desire to recreate their nation than any other action on our part."[83] Marshall Plan aid and the 1948 Berlin Airlift further assured the German population of American goodwill, as did American commitment to restoring Germany's industrial production and prosperity in the face of Russian and French objections. Moreover, countless millions of encounters between German citizens and their American occupiers convinced Germany of America's good intentions, despite the lingering bitterness of the war. Germany found in America the ally it had never found in Europe, and under American occupation it seized a new democratic and western self-identity. Even today, almost two decades after German reunification, Germany remains a strong ally, and American forces are still stationed in Germany, a direct consequence of America's generous postwar occupation.

The war in Europe was not just an occupation of Germany and Austria but also a liberation of much of western and southern Europe. The distinction between occupation of enemies and liberation of friends was central to civil affairs planning and execution in World War II. Under direction from Washington, Eisenhower had negotiated agreements with various military committees of the Allied governments in exile in London—that is, from countries that were occupied by the Nazis—which laid out how his advancing armies would assist in the restoration of their regimes and how the restored regimes would assist the Allies in the final assault into Germany.

The Franklin Roosevelt administration was careful, many thought overly so, to recognize only those regimes-in-exile that it believed truly represented the interests of the majority of the citizenry, thereby avoiding the de facto American legitimization of collaborationist, communist, or factional groups in the liberation process.

In the case of France, against the advice of the British and General Eisenhower, President Roosevelt refused to recognize Gen. Charles de Gaulle's French Committee for National Liberation as the sole governmental authority for France, mainly because de Gaulle's government had never been legitimated by elections within France.

Allied procedures and responsibilities for reestablishing popular governance in France after D-Day were therefore not negotiated directly with de Gaulle's government in Algiers but rather coordinated directly with lower-ranking French officers in London prior to the invasion.[84] Fortunately, goodwill, practical necessity, and similarity of interest among the Frenchmen and American soldiers on the ground in Normandy and farther into liberated France led to a well-executed liberation, despite the poor relationship between Roosevelt and de Gaulle at the top. Indeed, both parties were in substantial agreement on the main lines of operation: retreat of the German forces out of France, the disestablishment of the Vichy government, the restoration of French civilian control, and the speedy onward movement of the Allied armies into Germany itself.

Eisenhower ordered that "Civil Administration in all areas will be normally controlled by the French themselves."[85] As implemented by Patton's Third Army, this meant "any semblance of military government in France was to be scrupulously avoided, and the French would resume full civil activity as fast as conditions permitted."[86]

The Third Army after action report elaborates:

> In all echelons the spirit and letter of the Supreme Commander's order
> were felt as corresponding to the use of the official designation "Civil
> Affairs" rather than that of "Military Government," which was to obtain
> in Germany. From the first landing on the Norman coast, civil affairs
> officers were instructed to secure the cooperation of civilian police,
> transport, and communication authorities without infringing French
> rights or sensibilities, and to keep watch upon themselves to avoid all
> appearances of usurping the powers of military governors . . . The mayors
> at first were usually appointed by allied CAO's [civil affairs officers]
> in consultation with De Gaulliste liaison officers and they promptly
> recruited large special police forces, mainly from boys of the *Resistance*,
> to reinforce the *Gardes Champetres* and gendarmes in Normandy
> and Brittany.[87]

The U.S. Army's civil affairs teams deployed in the second wave of the amphibious assault on Normandy beach and the airborne attacks behind the Norman coast and rushed immediately to occupy the seats of local authority and ensure compliance with Allied military needs. In all, some one thousand civil affairs soldiers, including two hundred officers, accompanied the Normandy invasion.[88] The civil affairs teams included French liaison officers usually supplied by de Gaulle's Free French Army.

Irrespective of Roosevelt's desires, de Gaulle had also organized for the French liberation, ensuring that his loyalists inside France took the necessary actions to spread the Gaullist power base throughout the liberated areas.[89] Cooperating with Allied officers to the maximum extent possible was the Gaullist strategy for taking the reins of government in France by fiat. Indeed, as the U.S. Army broke out from Normandy beachheads and the Germans began to withdraw, the French still behind German lines began organizing themselves for their moment of liberation.

> As the Third U.S. Army moved forward, increasing strength and
> better organization in civil government were found. Orleans and the
> surrounding area were ably administered. The originally cordial relations
> with French officials were confirmed on August 25 by the "Revised
> Directive for Civil Affairs Operations in France," which made civil affairs
> policies conform with the agreements concluded by the United States
> and Great Britain with the provisional government and defined the
> powers of each in the zones of operation. In Lorraine, and particularly
> in Nancy, where the Third Army had its headquarters from October 1,
> 1944 to January 3, 1945, the record of efficiency and cooperation on the
> part of the French with the American forces was outstanding.[90]

* * *

Cooperation went both ways. The U.S. Army assisted the French in restoring utilities, communications, and transportation infrastructures, especially those useful to the war effort, as well as providing emergency supplies of food and public health items. The French, for their part, assisted the Allies with traffic control, police support, and other rear-area functions vital for carrying the offensive into Germany.

By October 1944 de Gaulle had captured so many of the provincial and municipal governments in France that Roosevelt had no option but to recognize his government. In essence, two parallel and complementary liberations had occurred, one by the Allied armies and their civil affairs teams and another by the Gaullists. The civil affairs approach to liberation would produce similar outcomes in Norway, Luxemburg, Belgium, the Netherlands, and many other liberated areas.

World War II: Japan

Japan's war of conquest in the Pacific was intended as a permanent occupation of its Greater East Asian Co-Prosperity Sphere, but only as a limited war against the United States. Japan's military leaders gambled that the country's surprise victories along America's Pacific frontier—Hawaii, the Philippines, Wake, Guam, Midway— would cause Americans to accept Japanese dominance in the western Pacific. Japanese strategy faced two insurmountable hurdles, the first being the country's inability to conquer the vastness of China, the second being America's resolve to crush Japan following Pearl Harbor.

America's decision to destroy the Japanese military, achieve unconditional surrender, and occupy Japan forced Japan into a costly defensive struggle that it ultimately could not win. From 1943 on, the United States imposed a merciless attrition warfare on Japan on land, sea, and air from a position of ever-increasing strength. In many categories of combat, American forces enjoyed a kill ratio advantage of over ten to one against their Japanese opponents. In the hugely unequal struggle that developed, the Japanese fanatically sacrificed their lives to deter the American invasion forces, but to virtually no effect.

By the summer of 1945, Japan's armed forces had been largely destroyed. Nearly 7 million soldiers, most of what still remained of the Japanese army, were stranded on the Asian mainland or on bypassed Pacific islands and were therefore unable to defend the mainland.[91] Japan's navy and merchant marine were nearly all sunk and rusting on the bottom of the ocean.

Though conventionally defenseless and near starvation, and with the Americans assembling the largest and most powerful invasion force in history for a final attack starting in the fall of 1945, the military in Japan still urged a protracted struggle on the home islands. However, the rapidly escalating B-29 firestorm attacks on Japanese cities and, finally, the dropping of atomic bombs put the population's very existence at risk.

Japan's World War II demographic experience mirrors Germany's. Out of a population of 70 million, Japan reached a peak mobilization of 6.1 million men. Of these, 2.6 million died in the war and another 326,000 were wounded (4 percent of population). Japan suffered 659,000 civilian deaths in the war (slightly less than 1 percent), disproportionately in the last months of the war, mostly from the bombing attacks and malnutrition. Forty percent of Japan's urban area was destroyed.[92] Close to 9 million Japanese civilians were homeless by war's end. Another 3.5 million civilians were trapped overseas, at the mercy of their Allied captors or their former colonial subjects.[93] Famine, malnutrition, and disease stalked the Japanese mainland.

General MacArthur, the Supreme Commander for the Allied Powers (SCAP), wrote, "Never in history had a nation and its people been more completely crushed than were the Japanese at the end of the war . . . Their entire faith in the Japanese way of life, cherished as invincible for many centuries, perished in the agony of their total defeat."[94]

Emperor Hirohito's August 15, 1945, radio address summoned the Japanese people to reconcile themselves to the nation's unconditional surrender:

TO MY GOOD AND LOYAL SUBJECTS:
After pondering deeply the general trends of the world and the actual conditions obtaining in Our Empire today, We have decided to effect a settlement of the present situation by resorting to an extraordinary measure.

We have ordered Our Government to communicate to the Governments of the United States, Great Britain, China and the Soviet Union that Our Empire accepts the provision of their Joint Declaration.

To strive for the common prosperity and happiness of all nations, as well as the security and well-being of Our subjects, is the solemn obligation which has been handed down by Our Imperial Ancestors, and which We lay close to heart. Indeed, We declared war on America and Britain out of Our sincere desire to ensure Japan's self-preservation and the stabilization of East Asia, it being far from Our thought either to infringe upon the sovereignty of other nations or to embark upon territorial aggrandizement. But now the war has lasted for nearly four years. Despite the best that has been done by everyone— the gallant fighting of military and naval forces, the diligence and assiduity of Our servants of the State and the devoted service of Our one hundred million people—the war situation has developed not necessarily to Japan's advantage, while the general trends of the world have all turned against her interest. Moreover, the enemy has begun to employ a new and most cruel bomb, the power of which to do damage is indeed incalculable, taking the toll of many innocent lives. Should We continue to fight, it would not only result in an ultimate collapse and obliteration of the Japanese nation, but also it would lead to the total extinction of human civilization. Such being the case, how are We to save the millions of Our subjects; or to atone Ourselves before the hallowed spirits of Our Imperial Ancestors? This is the reason why We have ordered the acceptance of the provisions of the Joint Declaration of the Powers.

We cannot but express the deepest sense of regret to our allied nations of East Asia, who have consistently co-operated with the Empire toward the emancipation of East Asia. The thought of those officers and men as well as others who have fallen in the fields of battle, those who died at their posts of duty, or those who met with untimely death and all their bereaved families, pains Our heart night and day. The welfare of the wounded and the war-sufferers, and of those who have lost their homes and livelihood, are the objects of Our profound solicitude. The hardships and sufferings to which Our nation is to be subject hereafter will certainly be great. We are keenly aware of the inmost feelings of all ye, Our subjects. However, it is according to the dictate of time and fate that We have resolved to pave the way for a grand peace for all the generations to come by enduring the unendurable and suffering what is insufferable.

Having been able to safeguard and maintain the structure of the Imperial State, We are always with ye, Our good and loyal subjects, relying upon your sincerity and integrity. Beware most strictly of any outbursts of emotion which may engender needless complications, or any fraternal contention and strife which may create confusion, lead ye astray and cause ye to lose the confidence of the world. Let the entire nation continue as one family from generation to generation, ever firm in its faith of the imperishableness of its divine land, and mindful of its heavy burden of responsibilities, and the long road before it. Unite your total strength to be devoted to the construction for the future. Cultivate the ways of rectitude; foster nobility of spirit; and work with resolution so as ye may enhance the innate glory of the Imperial State and keep pace with the progress of the world.[95]

As with Germany, the completeness of the Japanese defeat opened the country to reconstruction along lines suitable to both America and Japan. But while Clay was forced by circumstances to rebuild Germany from the bottom up, MacArthur occupied a country with a functioning national government and decided to take full advantage of that fact. Rather than disband the government and risk the anarchy that could occur, MacArthur would force the Japanese in power to admit their errors and reform themselves.

MacArthur found a fortunate and timely ally in the emperor himself, who on the occasion of the Japanese surrender, without discussion with MacArthur, publicly renounced the war, his divinity, and the superiority of the Japanese race.[96]

MacArthur used Hirohito's authority for his own purposes. Rather than issuing over his signature the three initial occupation proclamations drafted by his military governance staff, MacArthur instead had Hirohito issue them in the emperor's name directly to the Japanese government and people.[97] The Imperial Ordinances made it a crime, punishable in both the Japanese and U.S. military legal system, for any Japanese to disobey their occupiers.[98]

Although the atomic bombing of Hiroshima and Nagasaki accelerated V-J Day by perhaps a year by obviating the need for a ground invasion of the Japanese home islands, the planning for military occupation and reconstruction of Japan was already near completion in August 1945. Indeed, as with Germany, the planning for postwar had begun years earlier in the immediate aftermath of the Pearl Harbor attack. Beginning in February 1942, Secretary of State Cordell Hull chaired the President's Advisory Committee on Postwar Foreign Policy, a group that included high-ranking army, navy, and congressional leaders.[99] The advisory committee is notable for its recommendation of the unconditional surrender policy. Beginning in late 1943, interagency work, again led by State, was broadened to consider the postwar occupation policies in much greater detail.

In autumn 1944 the Interdivisional Area Committee on the Far East recommended positions on American postwar objectives with respect to Japan, including the duration and extent of the occupation, the composition of occupation forces, and various policies that the military government would enact.[100] Although the State Department had the lead, the coordination between State and the War Department's Civil Affairs Division was remarkable. Hugh Bolton, one of State's primary planners and later the head of the Japan Division, was a lecturer at the army's School of Military Government in Charlottesville beginning in 1942 and thereby had a direct hand in preparing the thousands of civil affairs personnel who would eventually serve in Japan under MacArthur.[101]

Beginning in January 1945, planning responsibilities were transferred to the State-War-Navy Coordination Committee (SWNCC), which decided the remaining issues and provided the final instructions to the Civil Affairs Division and General MacArthur. The SWNCC reaffirmed that the occupation would be three-phased: an initial period of stern military government to ensure demilitarization, a second period of "close surveillance" and progressive relaxation of controls to achieve democratization, and a final period of reintegration into the international community.[102]

In June 1945 the SWNCC began drafting final instructions for the Supreme Commander for the Allied Powers. Known as SWNCC 150, "Political-Military Problems in the Far East: United States Initial Post-Defeat Policy Relating to Japan," it directed Japan's permanent disarmament, retention of the emperor, a top-down approach for governing Japan through the intact national government, and numerous other policies. Revised as SWNCC 150/3 in the week following Japan's August 14 agreement to surrender, the document was hand-carried to MacArthur at his headquarters in Manila.[103] In anticipation of the atomic bomb making an imminent Japanese capitulation possible, the head of MacArthur's Military Government Section, Brig. Gen. William Christ, was recalled to Washington from late July to early August 1945 to finalize the basic plan for the military governance of Japan.[104]

By August 10, 1945, over two thousand military government officers from the army and navy plus an additional four thousand enlisted men had been specifically trained and organized for occupation duty in Japan. All of the officers and many of the enlisted men had been carefully selected for governance duty based on background, training, or technical specialty. Of this six-thousand-man governance force, more than a thousand officers were already forward stationed at the Civil Affairs Staging Area in Monterey, California.

During the week of August 18 to 25, 297 military government officers were flown from Monterey to Manila. Most of this initial group was assigned to the various subordinate occupation headquarters, and 75 were retained by MacArthur's Military Government Section. By September, 800 of the military government officers and 900 of the enlisted men were in theater, with the remaining thousands en route. By the time the force flow ended that fall, SCAP's Military Government Section would have over 450 officers and nearly 300 enlisted men assigned; Sixth Army headquarters and Eight Army headquarters, combined, would have 275 military government officers and 240 men assigned; and 800 officers and 2,000 enlisted men—far more than originally planned—would be assigned to XXIV Corps, which had responsibility for Okinawa and the increasingly tense situation in Korea. (Japan's complete cooperation with its occupiers made the diversion possible.)

The tactical echelon of the military government organization included provincial groups and provincial companies for local administration.[105] In design, the military governance structure in Japan mirrored the Japanese government's bureaucratic and prefecture structures and was responsible for ensuring that the orders issued by MacArthur to the Japanese government in Tokyo were faithfully transmitted to and executed by all governmental entities throughout Japan.

Washington may have provided the policy and the personnel, but MacArthur supplied the execution and the flair. Upon his arrival in Japan on August 30, in SCAP General Order Number 1, MacArthur directed the Japanese military commanders to immediately disarm and demobilize their 250,000 troops in and around Tokyo, which they did. MacArthur and his 3,500-man staff established themselves in what became known as the "Dai Ichi" (Number One) building, symbolically overlooking the Imperial palace, and supervised virtually every aspect of Japanese government.[106] Observers could not help but note how the Japanese people developed an intense loyalty toward MacArthur, whom they viewed as the man who would deliver them from the abyss they had created.[107] Cajoling the various Japanese ministries if possible, but directing them if necessary, the Supreme Commander Allied Powers and his staff thoroughly remade Japan along progressive lines: purging militants, instituting sweeping land reforms, guaranteeing civil liberties and free press, promoting women's rights and labor unions, liberalizing and expanding education, secularizing government, instituting a new Japanese constitution, and developing a parliamentary democracy.[108]

As in Germany, shiploads of U.S. food prevented mass starvation and created an atmosphere of goodwill. MacArthur once famously cabled Washington, "Give me bread or give me bullets," so important was food to postwar stability.[109]

Moreover, the innate kindness and generosity of the average GI led to even greater trust and receptiveness among the Japanese. MacArthur sensed the importance of the Japanese getting to know the Americans and refused to institute the "no fraternization" policy so notoriously in effect in Germany. He also ordered five-year prison sentences for American service members caught slapping a Japanese citizen.[110]

Initially, the occupation planners had estimated that 600,000 soldiers would be needed in Japan. As it turned out, the occupation force peaked at 354,000 in December 1945, was reduced to less than 200,000 by summer 1946, and kept declining until it reached 150,000 in 1948. The Japanese military, like the German military, was completely disbanded; however, the 94,000-man Japanese civil police force remained on duty, subject to American supervision.[111] As in Germany, mutual cooperation led to a relatively quick restoration of Japanese sovereignty. The first nationwide elections, and by far the most democratic in Japanese history, occurred in April 1946, when Japanese voters of both genders elected representatives to the Diet. The new draft constitution, a focus of the election, was thoroughly debated and ultimately ratified by the Diet in 1947.

Also in 1947, despite the new Japanese constitution's renunciation of warfare, MacArthur allowed Japan to rearm a small home defense force of seventy-five thousand service members; of course it was closely allied with the United States. Formally, the Allied occupation of Japan lasted until 1952. Since that time, America and Japan have stayed closely allied, and U.S. military forces retain bases in Japan and remain committed to Japanese defense.

In both its completeness and pace, the occupation of Japan was a surprising and unqualified success. America had reduced an alien culture to ashes and out of the ashes had built a transformed society. Within two years of the Japanese surrender, wise and generous occupation policies had largely changed the Japanese identity; where there had once been a semifeudal and xenophobic militaristic empire, there now appeared a democratic and receptive nation yearning to enjoy the blessings of peace. As the Japanese newspaper *Asahi* editorialized upon MacArthur's departure in 1951:

> We have lived with General MacArthur from the end of the war until today. . . . When the Japanese people faced the unprecedented situation of defeat, and fell into the *kyodatsu* condition of exhaustion and despair, it was General MacArthur who taught us the merits of democracy and pacifism and guided us with kindness along the bright path. As if pleased with his own children growing up, he took pleasure in the Japanese people, yesterday's enemy, walking step by step toward democracy, and kept encouraging us.[112]

A more successful conclusion to America's war with Japan can scarcely be imagined, nor could a greater compliment be paid to the American way of war as practiced at the time.

Before we leave our discussion of World War II, it is necessary to review some of the demographic data for some of the other combatants. The Soviet Union, which fought to defend its soil for the majority of the war, reached a peak mobilization of 12.5 million men, really never fully recovering from its enormous 1941 losses, and over the four years of the war suffered over 8.6 million military deaths, 14.6 million wounded, and 2.8 million prisoners of war who survived. Another 7 million to 12 million Soviet civilians died in the war.[113] Fully 10 percent of the Soviet prewar population of 196 million perished in World War II.

It is indeed remarkable that the Red Army concluded the war on the offensive, given the tremendous Soviet Union losses among military-age males, well over 50 percent in all categories. Indeed, by the end of the war the Red Army was running out of infantry and was actually growing weaker. By contrast, the United States reached a peak mobilization of 16.4 million, about 12 percent of the population, and suffered 407,000 deaths and 672,000 wounded. Britain reached a peak mobilization of 4.7 million, about 10 percent of the population, and also suffered about 400,000 military deaths. France lost fewer than 250,000 men, most of them in the forty-day blitzkrieg of 1940 that ended in the country's surrender, occupation, and collaboration.

Due to France's stunning incompetence in its prewar defensive strategy, Germany had finally achieved the complete victory over the French army that Clausewitz had wanted. But the nineteenth-century world of limited dynastic wars, Clausewitz's world, had long since ended, and World War II became an attrition-dominated total war in which the competing powers demanded complete, unconditional, world-changing victory. There would be no war-ending agreement between princes who defended each others' right to rule.

American Military History and Governance-Based Warfare Theory

So, how can we fairly summarize the American military experience through the end of World War II?

Most obviously, our successful offensive wars, meaning wars where we changed the governance of other nations, were almost always resolved through wars of attrition, not through battles of annihilation. In our successful offensive wars, the defender simply lost the demographic capacity to continue the war, mainly through a combination of very heavy losses (around 50 percent) in its military-age male population and some losses among its noncombatant population of women, children, and old men. Only when America had convincingly demonstrated the inability of the defender's military forces and their leadership to protect their population and preserve the status quo antebellum had we established the precondition that enabled the defeated nation to make the psychological transformation from combatant to vanquished.

America's successful offensive wars had always ended in a prolonged occupation, military governance, and gradual restoration of "supervised" sovereignty to the defeated nation. Every attempt was made to reconstruct the defeated nations according to liberal democratic values.

Conducting offensive wars in accordance with the rule of law and published general orders was instrumental to our successful occupations of Mexico, the Confederacy, the Philippines, Germany, and Japan. Indeed, from the Indian Removal Act and the beginnings of the reservation system through World War II, responsible army commanders devoted great forethought and energy to the legal codes and governance solutions that would be, over the course of the war, imposed on the nations we defeated. Enemy resistance to the army's imposed laws, whether through a defensive campaign of conventional military battles or through unconventional measures, merely delayed the inevitable "just" occupation and proffered governance.

Our offensive war policy has been similar to a "good cop-bad cop" routine, where we punish the defender, physically and psychologically, until he finally consents to his inevitable fate. When American soldiers have conducted themselves in accordance with published laws of warfare and imposed just, albeit firm, military governance over occupied territories, we have perhaps made it easier for our enemies to surrender themselves into our trust.

In our few defensive wars, where the continuance of the American government was at stake, the main strategy was to prolong the conflict, making it too costly for the attacker to continue to wage. Single decisive battles of annihilation never materialized. Rather, it was the cumulative effect of numerous indecisive battles that caused our attackers to lose the will or ability to continue their aggression.

In World War II the defensive war ended when the United States became the attacker and, in response, repaid Germany and Japan with our own offensive war and American occupation.

Wars of liberation, where we restored to a nation its legitimate government, required only modest military effort over and above the necessary force-on-force counteroffensive operations. However, economic and humanitarian assistance has always been useful in promoting stability in liberated lands. The litmus test for whether we are conducting an occupation or a liberation seems to be whether there is among the population substantial living memory of liberal traditions and democratic governance. Restoration of liberal government in the occupied countries of western Europe after World War II was rather easy and in post-World War II Italy slightly harder, but in Cuba, which had no memory of liberal local government, even prolonged military governance was unsuccessful in developing enduring democratic rule.

Our problem with the limited wars we have pursued, where our own governance or that of our adversary was not initially meant to be at stake, was keeping the wars' aims and means limited to the specific issues at hand. We were successful only in the Spanish-American War. In the War of 1812, in the Mexican War, and, as we shall see later, in Iraq, limited wars escalated to the point that one government or the other was either overthrown or threatened with destruction.

It must also be said that Americans have always been especially good at force-on-force attrition warfare, a pattern that continued through the Cold War conflicts and persists to this day. Americans nearly always win the firefight. Historically we have been unequalled by any foreign opponent, and the soldiers and marines now fighting the numerous battles and engagements in Iraq and Afghanistan continue the tradition of martial excellence that has served the nation well for centuries. When the decisive battle and the quick victory have become a faded dream, American soldiers have historically proven and continue to prove our ability to achieve kill ratios far higher than any opposing army and eventually bring it to surrender. Since the Civil War, American war deaths never reached even one-third of 1 percent of the American population, well below the historic demographic benchmarks seen in our early wars.

Our ability to win wars of attrition is not unique to us; many nations throughout history have fought and won such wars. However, the United States is exceptional in that we have demonstrated the ability to successfully defeat nations and then occupy and remake them along liberal lines that are beneficial to America and to the vanquished people and, indeed, to make them friends and allies. This truly remarkable combination of war fighting and peace making is rarely seen in all of history. In fact, when we consider the American experience, growing in little more than three hundred years from a few hundred colonists in Plymouth and Jamestown to the greatest superpower ever known, we can only be completely impressed by the unequalled military achievement of the American people.

It is especially galling, therefore, that we now fail to recognize the reasons for our historic success in our current war-fighting doctrine but rather embrace the transient knowledge of the nineteenth-century foreign philosopher Clausewitz, who came from a military tradition much less successful than ours. He could learn more from us than we could ever learn from him.

Chapter 4

American Conflicts During
the Cold War Era:
The Unraveling of the American
Way of Offensive War

The successful American offensives that brought Allied victory in World War II were the most spectacular overseas wars ever conducted by any nation in recorded history. But they didn't change the world as much as the occupations of Germany and Japan that followed. The American government was adamant not to repeat the errors that muddled the World War I armistice, when the Central Powers were allowed continued self-governance and, in their bitterness over defeat, rearmed for a second, larger war. America, no longer a junior partner in Allied postwar schemes, demanded that the new peace permanently resolve the problem of German and Japanese militarism and very early in the occupations concluded that democratization of Germany and Japan along American lines was likely to be a feasible and desirable means to make them willing participants in America's new international system. The occupation and westernization of Germany, Japan, Korea, and many other U.S.-occupied or liberated nations would monopolize the army's overseas efforts in the immediate postwar period. Consolidating the World War II victory was task number one.

A handful of Americans, a few in positions of responsibility, advocated immediate war with the Soviet Union, which, though an ally since 1941, by war's end was clearly working to establish a competing empire based on its inimical Communist totalitarianism. True, both sides tested each other along their line of contact, which generally was their ground forces' limits of advance, as established in the Yalta agreements. But none of the provocations was thought to be sufficient reason for war. The overwhelming concern of the American people in the immediate postwar period was demobilizing its 12 million service personnel, many of whom had been overseas for years, and restoring them to their families and their careers as quickly as possible.

As the elected representatives of a middle-class democracy, American politicians had little inclination to plunge the nation into a new "elective" war against our ally, merely on the suspicion that he may someday become an enemy. Preemptive war was not part of the American military tradition at that time. Americans viewed themselves as a peaceful people and preferred that the other side attack first. The American pattern was to suffer the first blow, and then, riding a wave of popular indignation and righteousness, crush the nation that had broken the peace and made itself our enemy. The blame for the war would not be ours, but theirs. Owning the moral high ground was viewed as more important than winning the first engagement.

So the Cold War became America's first defensive struggle since our formative struggles against England in the Revolution and the campaigns of 1814 and 1815. From 1945 to 1989, America sought only to preserve what we had won in World War II. The idea was to contain the Soviet Union within its World War II occupation zones and to strengthen and defend democracy in ours, in both Europe and the Far East.

The Cold War achieved its military aims: not a single nation on the American side of Russia's iron curtain fell to the Red Army or its surrogates in the next forty-five years, though many areas were militarily tested—Greece, Berlin, and Korea. The Soviets were more successful in aligning themselves with anti-imperialist movements in Vietnam, Cuba, and elsewhere, but again American defensive efforts against emerging Communist threats generally proved successful. As enunciated in George Kennan's famous "X Article" in *Foreign Affairs*, "The main element of any United States policy toward the Soviet Union must be that of a long-term, patient but firm and vigilant containment of Russian expansive tendencies. . . . Soviet pressure against the free institutions of the Western world" must be countered through "adroit and vigilant application of counter-force at a series of constantly shifting geographical and political points, corresponding to the shifts and maneuvers of Soviet policy." Such a policy, Kennan predicted, would "promote tendencies which must eventually find their outlet in either the break-up or the gradual mellowing of Soviet power."[1] A strategic defense for an extended time would further American interests, so defense became the American postwar grand strategy. Accordingly, as the World War II generation of American soldiers died off, they would be replaced by generations of soldiers unaware of the demands of offensive war.

The Korean War

Korea was occupied and colonized by the Japanese in 1910. In accordance with World War II agreements among the Allies, the U.S. military would occupy the Korean peninsula south of the thirty-eighth parallel upon the Japanese surrender, while the Soviet Red Army would occupy the northern half of the Korean nation.

Remarkably, the U.S. military was completely unprepared to occupy southern Korea when the Japanese announced their surrender in August 1945. With Washington's focus clearly on Japan and Germany, Korea was a low priority, to the point that the U.S. Army's military government schools were not allowed to even study the Korean problem until the fall of 1945.[2] Consequently, the Americans could not even assess whether they were about to "occupy" a hostile Japanese protectorate or "liberate" a friendly people from Japanese rule. Hongul speakers were simply unavailable.

The task of occupying Japan fell at the last minute to the XXIV Corps on Okinawa, veterans of the ugliest fight in the Pacific. Their commander, Lt. Gen. John R. Hodge, was completely unprepared for the mission, having no personal knowledge of Korea and no experience with military government. His staff was no different. Hodge delegated the military government mission to his antiaircraft artillery commander. This unlikely command team was soon augmented by twenty experienced military government officers who had been trained for duty in Japan; none had any knowledge of Korea.

So when American forces finally arrived in southern Korea on September 8, 1945, and proclaimed military government the following day, they did so in nearly complete ignorance of what they would find and what they would do.[3] Needless to say, there were a great many missteps early on. According to many observers, the snap decision to occupy and militarily govern was a root of all evil; liberation and a civil affairs approach, similar to that used in France, would have been more appropriate to the situation.[4]

Having ruled Korea for thirty-five years, the Japanese and their Korean collaborators were firmly entrenched in all aspects of government and commerce. Resurgent Korean nationalism demanded that the hundreds of thousands of Japanese be deported immediately and their Korean henchmen be quickly removed from authority.[5] Hodge accommodated as quickly as he felt practicable, though not nearly as fast as the Koreans desired. The result was economic decline and instability. Though Korea was largely unscathed by the war, unlike Germany and Japan, living standards actually fell in American-controlled Korea, and Hodge seemed unable to craft assistance programs and economic policies to turn the situation around. Reflecting the gloom, in March 1947 John Hilldring, the assistant secretary of state, reported, "Many Koreans feel they are worse off than they were under the Japanese."[6] Lacking the political savvy and liberal inclinations of MacArthur and Clay, Hodge, ever the combat soldier, could not effectively tutor the Koreans in democratic politics and procedures, and reflexively supported right-wing anti-Communist factions. Indeed, as the American occupation ended in 1948, South Korea was politically unstable, socially chaotic, and economically bankrupt. The Seoul reporter for the *Chicago Daily News* wrote at the time, "An atmosphere of defeatism and frustration grips Americans in this dreary, dusty capital as they prepare to withdraw after three years that many of them feel has been wasted."[7]

Although neither the Soviets nor the Americans had envisioned a permanently divided Korea during their World War II conferences, the incompatibility of the emerging Communist governance in the North and the more western-looking governance in the South worked against unification of the two occupation zones. Syngman Rhee, a bona fide Korean nationalist, with a family lineage linking him to the precolonial Korean monarchy and a portfolio that included the presidency of the anti-Japanese Korean government in exile in Shanghai from 1919 to 1925, quickly emerged as the leading politician in the South. American educated and Christian, Rhee was an acceptable choice to the American military authorities in Seoul, despite his authoritarian tendencies, and he was elected president of South Korea in 1948, the same year that America abandoned hopes of reuniting the occupation zones and granted South Korea its sovereignty.

Upon Korean independence, the U.S. military released to Rhee's command the very large constabulary of Korean soldiers that America had created during the occupation, redesignated as the Army of the Republic of Korea (ROK). America retained a sizable advisor corps with the ROK. In reaction, the North Koreans under Kim Il Sung established an anti-Rhee Communist South Korean Labor Party, which by 1949 grew to "several thousand guerillas backed by 10,000 party members, 600,000 active sympathizers and up to 2 million 'fellow travelers' in affiliated front organizations."[8]

Guerilla warfare ensued, first on Cheju-do Island in 1948 and later on the mainland. The ROK brutally suppressed these rebellions using the harshest of hard-war measures, which were modeled on Japanese practices: forced relocations of peasants, cordon and search, often indiscriminate reprisals, mass arrests, and scorched-earth destruction. Fearing that the ROK would alienate the populace, American advisors pressured the Rhee government into adopting a "half force, half administration" counterinsurgency strategy, featuring less aggressive military action and made more palatable by land reform, civic improvements, and amnesty programs. By the end of 1949, the counterinsurgency war was at least on track, though not won.[9]

In June 1950 North Korea launched an offensive war to occupy South Korea, destroy Rhee's government, and reunify the Fatherland. America responded by sending conventional forces to assist in South Korea's defense. Some in the American camp, especially the American and United Nations commander,

Gen. Douglas MacArthur, desired to expand our war aims to include the occupation and remaking of North Korea, in keeping with our World War II practice of eliminating aggressive regimes. After a summer of unexpectedly hard fighting, American forces defeated the North Korean attack with the stunning September 15, 1950, Inchon landings and coordinated counterattacks from the Pusan perimeter.

With the success of the Inchon landings, the breakout from Pusan, and the subsequent recapture of Seoul, American combat forces streamed north with American military governance teams in tow, intent on the occupation and political reconstruction of the Communist North.[10] By October American military government units from the United States were moving to Korea to capitalize on the sudden collapse of the North Korean army and help in the occupation of the North.[11]

MacArthur, however, did not envision a U.S. military government of North Korea, perhaps recalling the problems of the Hodge years, and wanted Syngman Rhee's government and the ROK army to take the lead in the political reunification of the peninsula. Indeed, ROK military government teams advanced with the United Nations (UN) forces, proclaimed military government in conquered areas, and began to bring the populations under Seoul's control.

Unfortunately, the reunification was not to be; in November 1950 the Chinese, until now observers of the war, successfully counterattacked south from the Yalu River, throwing the American-dominated UN forces back south of the thirty-eighth parallel, indeed south of Seoul, where the Chinese attack finally succumbed to withering American firepower.

None of the offensives in the war's opening months—the North Korean, American, or Chinese—achieved their hoped-for decisive victory, militarily or politically. Similarly, the offensives and counteroffensives in 1951 and 1952 tended only to inflict more casualties and bring the battle lines back closer to the thirty-eighth parallel. As the war of attrition stalemated, the war aims of all the belligerents became much more modest, essentially a restoration of the status quo antebellum. With all sides now playing defense, peace negotiations began.

The counterinsurgency struggle, however, continued in the South throughout the war, consuming considerable ROK strength and forcing the American military to devote significant effort to secure its rear area behind the front and its lines of communication. Hard-war tactics continued. In the end, the insurgency collapsed from heavy casualties, constant military and police offensives, lack of food and other support, and its inability to attract new recruits from the population in the South following the North Korean army's retreat in the fall of 1950.

When the armistice of 1953 finally ended the war, all of the prewar governments remained in power, although it is clear that both the South Korean and North Korean governments would have collapsed in 1950 had not the United

States and China, respectively, come to their assistance. North Korean losses were extreme, reaching the limits of the demographic envelope, with military casualties estimated at 215,000 killed, 304,000 wounded, 102,000 missing in action, and civilian deaths of perhaps a million, all out of a prewar population of 9.4 million.[12] Only Chinese intervention kept North Korea in the war. South Korea lost 1 million dead out of a population of 20.8 million (about 5 percent). Another 7.5 million South Koreans were made refugees or destitute.[13] The Chinese army suffered over 900,000 casualties. As in the Pacific War against Japan, the American military proved itself once again to be the gold standard of attrition war, achieving kill ratios over the Communist military, whether Chinese or North Korean, of about ten to one.

The Vietnam War

Many Americans view the Vietnam War as a defense of the independent South Vietnam against the aggression of the Communist North, but that view is skewed by our Cold War perspective that the United States was defending itself against worldwide Communist expansion. Although America's global defensive concerns were certainly valid, especially in light of Soviet offensive actions in Europe and Korea, from the Vietnamese perspective America was the aggressor, trying to impose an unpopular and unwanted government on the Vietnamese after the Allied defeat of the Japanese in 1945. Importantly, Ho Chi Minh's Viet Minh army had fought Japanese occupation of Vietnam during World War II, but the collaborationist French colonial regime had not. Roosevelt, who thought French colonialism to be overly exploitative, nevertheless deferred the decision on the ultimate postwar governance of Indochina. At the State Department, the Far East division urged Indochinese autonomy, while the Europeanists supported continuing French colonialism. Amidst Washington's indecision, the British, who were anxious that their own liberated colonies be returned to the Empire, maneuvered President Harry Truman into supporting European colonial restoration in Asia, including French Indochina.[14]

As Japanese power in Indochina collapsed in the summer of 1945, Ho took advantage of the anarchy and launched the August Revolution, taking control of the country, declaring Vietnamese independence, and securing the abdication and support of Emperor Bao Dai.[15] Ho, and his manifest popular support, was virtually ignored by the Americans and our allies, who in short order reoccupied Indochina and reimposed French colonial rule. In the thirty-year war that followed, the Vietnamese Communist understanding of the war never wavered. "They see the war entirely as one of defense of their country against the invading Americans, who, in turn, are seen merely as successors to the French."[16]

The French secured American support in Vietnam by pledging support for American aims in Europe, and America began providing some 80 percent of France's ever-escalating Vietnam War costs. After Mao's army reached the

China-Vietnam border in 1949, Ho gained a powerful source of material for his war effort and was able to escalate in turn.

The climactic French defeat at Dien Bien Phu in 1954, which cost the French fifteen thousand men, caused France to agree to a Geneva conference to end its colonial commitment, at least in the North. To emphasize that the South was also his, Ho had his forces destroy a French mechanized brigade, Group Mobile 100, in the Central Highlands as well. Ho went to Geneva as the victorious champion of Vietnamese nationalism, fully expecting that he and the Viet Minh would be given control of all of Vietnam.

In Geneva, the French conceded North Vietnam to the Communists and promised Ho nationwide elections in 1956 to reunify the country. Ho, sure he would win any fair election and sold out by his Russian and Chinese supporters, accepted the deal. However, America blocked the Geneva plan by installing a narrowly based, autocratic regime in the South under the leadership of Ngo Dinh Diem, a Catholic virtually unknown in predominantly Buddhist southern Vietnam but supported by powerful American political interests.[17] It was America's imposition of the Diem regime (and its successor Ky and Thieu regimes) on the Vietnamese population against its will, and against the spirit of the Geneva Accords, that caused the American war in Vietnam.

At first, we believed we could win the governance contest with the minimum of effort—a few hundred advisors and whatever billions of dollars of foreign aid. Diem, with Washington's concurrence, cancelled the 1956 elections required by the Geneva Accords based on assessments that Ho would win 80 percent of the vote.[18] But after the cancellation of the 1956 elections, the Viet Cong (VC), the reconstructed Viet Minh, organized in the South to topple the unpopular regime. American hopes that the American-equipped and -advised Army of the Republic of Vietnam (ARVN) would defeat the Communists on the battlefield proved overly optimistic, as did the hopes that the Diem regime would broaden its base of support among the common people. At its core, South Vietnam was an American creation built upon the remnants of French colonialism that could never inspire the willing sacrifice of ARVN's draftee soldiers. One of Diem's aides admitted, "Except for the color of our skin, we are no different from the French."[19]

During the Kennedy administration, the growing Viet Cong success and signs of imminent South Vietnamese collapse caused great strengthening of American resolve and expansion of U.S. effort. The war had to be won, even if not on the cheap as initially promised. Finally, in 1965, the Lyndon Johnson administration committed American infantry divisions to the fight, as did the North Vietnamese Army (NVA). The resulting attrition war, building for years, would determine the political fate of South Vietnam.

At the height of the war in 1968, American troop strength in Vietnam reached over half a million men, and two hundred to three hundred Americans were dying each week.[20] (The Communists, by comparison, lost ninety thousand men in battle in 1968.) The human cost of the war, though not demographically significant to a nation as populous as the United States, was prominently displayed on television, increasing its emotional effect. The political decision made in the 1968 presidential elections forced America into de-escalation of its Vietnam commitment, and over the next three years almost all U.S. ground troops were withdrawn.

As a salve, the United States attempted to strengthen ARVN to the point that it could maintain an independent South even without U.S. combat troops. In its peak year of 1972, ARVN forces would reach over a half million regulars, including naval and air forces, and 670,000 paramilitary and police auxiliaries. In March of that same year, combined NVA and VC forces in the South would number some 243,000, a small fraction of the South's totals. Though numerically and logistically inferior to ARVN, the Communists nevertheless launched a general offensive that most observers believed would have collapsed the South Vietnamese government had not American air power decimated the massed NVA ranks, which suffered about 50 percent casualties.

Despite the casualties, in 1975 North Vietnam committed 178,000 fresh troops to a new general offensive and in a matter of weeks routed the ARVN and captured Saigon to end the war. President Richard Nixon observed the failure of his Vietnamization strategy: "The real problem is that the enemy is willing to sacrifice in order to win, while the South Vietnamese simply aren't willing to pay that much of a price to avoid losing."[21]

The Vietnam War is remarkable not because of the American capitulation; initiators of offensive war have often decided to cut their costs and negotiate their way out of a conflict they started. Nor is the South Vietnamese defeat especially noteworthy; halfhearted efforts are rarely rewarded in war. Rather it is the Communist Vietnamese ardor in liberating their country from the foreign aggressor and, in their terms, its puppet regime that tested the limits of military commitment. North Vietnam mobilized 3 million men out of a population base of twenty million (in 1970), and the Viet Cong mobilized another million from the South. Of these, 1 million died and perhaps another million were wounded. North Vietnam's young men would unsmilingly joke that they were raised in the North to die in the South. The North also suffered from intense American bombing aimed at military and economic targets, though not population centers, resulting in 65,000 civilian deaths. No one knows how many VC sympathizers were among the 522,000 South Vietnamese civilians killed in the war, but certainly there were many. It would be reasonable to estimate North Vietnamese deaths at something less than 4 percent of the population and South Vietnamese Communist death rates, both military and civilian, at perhaps twice that of the North. Still, none of this death and destruction deterred the Communists. Ho once stated, "You can kill ten of my men for every one I kill of yours. But even at those odds, you will lose and I will win."[22] Ho had found the only military formula that could bring victory against America's overwhelming firepower efficiency.

Vietnam codified a new collage of security concepts that find their echo in this century's global war on terror: the belief that democratization and liberalization of the lesser-developed world will pave the way to peace; the belief that unpopular or narrowly based governments installed by the United States can be stabilized by appropriate military and political policies—that is, counterinsurgency operations; the belief that the governments we support will, over time, democratize and expand their popular support, allowing the U.S. military to hand over security responsibilities to the host nation; the belief that an interagency effort, rather than an army-commanded effort, is the best path to success; and the hope that the American electorate will patiently and lavishly support the stabilization process through multiple election years and economic cycles. This collection of interdependent propositions failed its test in Vietnam.

Structural reasons were at least partly to blame, because the host nation construct, with its sovereign government and independent military, creates power and policy interests independent of American control. To make matters worse, bureaucratic infighting between the U.S. military, the State Department, the Central Intelligence Agency (CIA), and the other departments of the "interagency" originated in the corridors of power in Washington, reverberated through the Military Assistance Command-Vietnam (MACV) headquarters and the U.S. embassy in Saigon, and played out in the villages of rural South Vietnam. For the first time in its history, the U.S. military fought an overseas war without unified military or political command and "lost" the war more for the failings of others than for its own.

Perhaps the army's greatest failing in Vietnam was not insisting on unity of command to ensure unity of effort. For instance, in October 1965 Gen. William Westmoreland announced that the U.S. Agency for International Development (USAID), whose director reports to the secretary of state, had "the primary U.S. Civil Affairs responsibility in RVN," and that army civil affairs units should confine themselves to support of tactical operations.[23]

To expand on President Nixon's remark, South Vietnam fell not because the army's counterguerilla operations failed to find and kill Viet Cong, or because the army failed to destroy NVA units that entered South Vietnam; all evidence suggests that the army succeeded in its combat missions. Rather, the war was lost politically by the South Vietnamese themselves, who never rallied sufficiently behind the cause of a free and independent South Vietnam under the governance of the Saigon-based elites. Extending the government's power base into the countryside was called "pacification" or, as President Johnson termed it, "the other war." From the very beginning of the struggle, pacification efforts suffered from a lack of intergovernmental (between Saigon and Washington), interagency (within both the South Vietnamese and U.S. governments), and intermilitary (ARVN and MACV) coordination, with no single agency having clear overall responsibility, while at the same time no agency was satisfied with the efforts that it felt were the responsibilities of the others.[24] The American interagency squabbling became so intense and counterproductive that in 1967 President Johnson briefly considered unifying all U.S. efforts in Vietnam under General Westmoreland along the lines of authority that MacArthur had enjoyed in occupied Japan, though by this late date investing in Westmoreland proconsulship over the government of South Vietnam was not deemed feasible.[25]

The compromise was Johnson's mid-1967 establishment of a deputy commander for Civil Operations and Revolutionary Development Support (CORDS) in MACV under Westmoreland. President Johnson filled the position with Ambassador Robert W. Komer, a CIA civilian who had served on the National Security Council and would leave Vietnam in 1968 to become ambassador to Turkey, and later Ambassador William Colby, also with a CIA pedigree. Civil Operations and Revolutionary Development Support oversaw not only counter-VC political operations such as the Phoenix Program but virtually all civil-military activities in the countryside.

> The breadth of CORDS programs was apparent from a listing of the programs and the agencies formerly charged with them: New Life Development (AID) [Agency for International Development], CHIEU Hot (AID), Revolutionary Development Cadre (CIA), Montagnard Cadre (CIA), Census Grievance (CIA), Regional and Popular Forces (MACV), Refugees (AID), Field Psychological Operations (Joint U.S. Public Affairs Office), Public Safety (AID), U.S. Forces Civic Action and Civil Affairs (MACV), Revolutionary Development Reports and Evaluations (all agencies), and Revolutionary Development Field Inspection (all agencies). CORDS also assumed coordination responsibility for pacification-related programs of the Agency for International Development, such as rural electrification, hamlet schools, rural health, village-hamlet administrative training, agricultural affairs, and public works. With few exceptions, all American programs outside of Saigon, excluding American and South Vietnamese regular military forces and clandestine CIA operations, came under the operational control of CORDS . . . An important exception was land reform, which the Agency for International Development insisted on retaining.[26]

The CORDS organization did not solve all U.S. interagency bickering, but it did allow for at least a coordinated effort between the military and the various civilian agencies. It effectively infused large numbers of highly qualified civilians into the pacification effort. In the I Corps area of northern South Vietnam, for instance, CORDS civilian personnel outnumbered CORDS military personnel by a margin of 1,250 to 750, despite the constant combat and danger. The State Department assigned several hundred foreign service officers to serve on CORDS provincial and district advisory teams, with individual tours ranging from eighteen to twenty-four months.[27]

After the 1968 Tet Offensive and prior to the American pullout, pacification efforts showed significant results. The South Vietnamese were sufficiently impressed that in 1969 they developed a similar system, known as the Central Pacification and Development Council, through which to coordinate the work of their ministries with the American effort.

Although much good work was done in the early 1970s, after the American military left—taking the CORDS effort home with it—none of it was sustained by the South Vietnamese themselves. Two decades of massive American military and civilian efforts could not impart lasting strength and legitimacy to the Saigon regime that we had created in 1954. In the final analysis, South Vietnam had become a "tyranny of the weak," the South Vietnamese government constantly threatening the United States with their imminent collapse if we did not give them everything they wanted. American aid for a time became virtually unconditional.[28] When the aid was scaled back to more reasonable and sustainable levels from the U.S. perspective, our South Vietnam experiment collapsed like a house of cards.

The Cold War in Europe

The Cold War, from the American standpoint, was purely a defensive, deterrent struggle aimed at preserving the democratic regimes that fought with us as Allies in World War II or emerged as friends in the war's aftermath. The long-term containment of Soviet expansionist tendencies was the American goal, not the rolling back of Soviet fait accompli in Eastern Europe or the removal of the Communist government in Moscow. Consequently, the U.S. military never procured the force structure necessary for an offensive war against the Warsaw Pact.

From an army standpoint, the Cold War was a forty-year preoccupation with preparation for a strictly defensive war. North Atlantic Treaty Organization planners would not consider offensive war scenarios, or even defensive war scenarios that resulted in counteroffensives into Warsaw Pact territory. United States Army thinking about offensive war, especially the necessary duties that accompany occupying conquered territory, atrophied. Army doctrine focused almost exclusively on defensive and counteroffensive operations designed to destroy attacking Warsaw Pact forces in the field. In many ways, Desert Shield and Desert Storm reflected the army's 1980s preparations for war in Germany: a

The end of the Cold War also illustrated the folly of illiberal occupation and the capacity of countries to reestablish their traditional forms of governance if only provided an opportunity. As the Soviet empire in Eastern Europe collapsed, followed by the breakup of the Soviet Union itself, nations that had long been occupied by the Red Army and had had Soviet-style governments forced on them for decades declared their independence from Russia and, for the most part, reestablished more liberal and representative governments based on their national traditions.

If the Russians viewed themselves as the beneficent liberators of Nazi-occupied lands, they soon became seen by the "liberated" populations as conquerors and occupiers who had no interest in allowing truly representative popular government anywhere. They provided security and stability for those who didn't oppose them, but never liberty or local sovereignty. For this they were despised and shown the door when the time came, perhaps never to be welcomed back. The Russians, and the Nazis before them, will always be seen as the worst occupiers imaginable. Occupation without the offer of national renewal is, one would hope, impossible to sustain.

Grenada 1983 (Urgent Fury)

Grenada, an island of only eighty thousand inhabitants located in the southern Caribbean, is an independent constitutional monarchy within the British Commonwealth. In 1979, Maurice Bishop and his New Jewel Movement overthrew the Grenadian prime minister, Sir Eric Gairy, and forged a strategic partnership with Fidel Castro's Cuba, inviting Cuban soldiers and engineers to begin construction of a nine-thousand-foot-long military airfield on the island.

Though initially arrested by Bishop during the coup, the Queen's appointed governor-general, Sir Paul Scoon, was soon released from jail and allowed to continue in office with reduced duties. In 1983, one of Bishop's colleagues, Dr. Bernard Coard, overthrew Bishop and imprisoned Scoon. This putsch led to protests, yet another coup by the military chief Hudson Austin, and the killing of Bishop, all over the course of a few days.

In league with the Organization of Eastern Caribbean States and several unaffiliated island governments, President Reagan ordered U.S. armed forces to go to Grenada, stop the mayhem, stop the Cubans, and restore the rightful government. On October 25, 1983, approximately seven thousand American soldiers, marines, airmen, and sailors invaded the island. In sporadic fighting that lasted a few days, fewer than a hundred Grenadians, Cubans, and Americans, combined, were killed.[36] Scoon was immediately released from house arrest and restored to his position as governor-general, a post he retained until 1992. Restored democratic governance is his legacy. The Cubans were rounded up and sent home in the immediate aftermath of the American invasion, and all U.S. forces had redeployed from Grenada by December. October 25th is celebrated as Thanksgiving Day in Grenada in commemoration of the U.S.-led liberation.

Panama 1989 (Just Cause)

The American military conducted another Caribbean liberation in December 1989 when it ousted the drug-trafficking Panamanian military leader Gen. Manuel Noriega and replaced him with President Guillermo Endara. As background, Endara had won the May 1989 election by a three-to-one majority over the Noriega candidate. In response, Noriega counterfeited the election results and had members of his "Dignity Battalions" publicly beat Endara, forcing him into exile in the American-occupied Panama Canal Zone. The Organization of American States immediately condemned the election processes and abuses.

In October several officers of the Panamanian Defense Force (PDF) tried to overthrow Noriega, even capturing him for several hours, only to be defeated and killed by Noriega's supporters.[37] Then Noriega began harassing and threatening American servicemen and family members of the U.S. Southern Command as they ventured out of the canal zone, eventually killing a marine lieutenant in Panama City on December 16, 1989.[38]

It was the last straw; enough was enough. Citing the need to stop Noriega's drug trafficking, safeguard democracy and human rights, protect American lives, and ensure the neutrality of the Panama Canal, President George H. W. Bush ordered some fifty-seven thousand U.S service members, many already stationed in Panama, to invade the country and overthrow Noriega. The long-planned and well-executed military operation commenced on December 20 and completely overwhelmed the much less capable pro-Noriega forces. Most of the PDF stayed in their barracks. In the week of fighting, twenty-three Americans and a few hundred Panamanians lost their lives.[39] Noriega remained on the loose until January 4, 1990, when he surrendered and was flown to Miami to await trial for his crimes. (He was tried and convicted in 1992.) President Endara was sworn in as president on December 20, 1989, just prior to the invasion, and remained in that office until 1994. A *CBS News* poll conducted in early January 1990 found that 92 percent of Panamanians supported the American military action and 76 percent wished we had sent troops earlier to support the October coup attempt.[40]

The Cold War in Retrospect

By the time the Soviet Union collapsed in the early 1990s, the U.S. military had forgotten all it had once known about offensive war. For forty-five years America had been playing defense. No soldier who had served in a high position during our World War II counteroffensive or in the postwar occupations was still alive. Even the privates and lieutenants who had participated in those campaigns and had made the army a career were long since retired from active duty. Army doctrine and training had meandered to a very limited comfort spot centered on winning the conventional force-on-force defensive battle on the plains of Central Europe, severely outnumbered, using a professional, high-tech army. Other missions, such as offensive warfare, occupation, and counterinsurgency, became fleeting chimeras at the edges of the army's collective peripheral vision, hardly seen or acknowledged. No officer with ambition would lay down his marker to force them center focus. And, perhaps most damning, the army as an institution could not find some way—for example, establishing a college or think tank studying all forms of war—to keep knowledge of such operations current in case they would one day be needed.

Still, at the end of the Cold War, there was no force on Earth with the power to expose the army's dereliction of its intellectual duty. The New Jewel Movement and the PDF were no match. In our first post-Cold War conflict, Desert Storm in Kuwait, Saddam Hussein's massive Third-World army proved itself a paper tiger, unable in any respect to meaningfully contest our forces. America, at its military zenith in the 1990s, had no rival. Among the world's powers, only America survived the conflicts of the twentieth century with its military power intact, indeed growing. Having enabled the rise of America to "sole superpower" status, the U.S. Army in the immediate post–Cold War period had earned the right to boast. Perhaps the time could have been better spent in preparing for upcoming challenges, but in the national celebration that was the 1990s, no one wished to consider that harder times might be ahead.

Chapter 5

Wars Not Directly Involving the United States

Offensive wars generally become wars of attrition by choice of the defender. The attacker may desire a cheap and quick victory, but the defender will resist the attacker as long as he is demographically able to continue fighting. Though the attacker may seek a relatively easy regime change, the defender, protecting his accustomed form of governance, will decide when to finally accede to the occupier's imposed governance. Should the attacker defeat the defender's conventional defense and occupy the land, the defender still may resort to a protracted struggle—guerilla warfare, terrorism, or civil disobedience—to force the attacker to withdraw. Making the cost of the occupation unbearable becomes the defender's goal. Increasing defender casualties, both military and civilian, to the point at which the defender submits becomes the attacker's only remedy.

As discussed previously, this has been the historical pattern in American warfare. But is this pattern specific only to us, reflecting uniquely American practices and attitudes, or is it more generally applicable? My admittedly brief survey of foreign wars shows that it is. Certainly we have seen the same approach used by Europeans and East Asians—namely the Japanese, Chinese, Vietnamese, and Filipinos—in our previous discussions of American wars. This chapter highlights four additional examples from foreign offensive wars of occupation.

First it is important to understand that examples of successful modern-era occupations are rare, which is probably the best evidence of their difficulty. German occupations during World War II, relying as they did largely on state-sponsored terror and coercive measures, all failed. Similarly, the Soviet occupation of Eastern Europe after World War II, also dependent on police-state methods, ultimately failed. The Soviets abandoned their conquests in the 1990s as the newly freed nations of the East turned to the West to defend their renascent liberal democratic identities. The rollback of Soviet socialism not only stripped Russia of its World War II acquisitions, it accelerated into a dismemberment of the USSR herself as nations that it had acquired during the imperialist expansion of tsarist Russia and the revolutionary wars of the early Soviet era declared their independence and set out on paths of their own. Lest we think that only totalitarian expansion schemes fail, following World War II the western European countries also abandoned their colonial possessions when their governance regimes were attacked by nationalists desiring independence. There is surprisingly little in the twentieth-century record, outside of America's achievements, to suggest that offensive wars can be "permanently" won. Only America seems to have consistently demonstrated the capacity to bring a country to its knees through attrition warfare and then rebuild it into a willing and content ally.

The Second Boer War (1899–1902)

At the same time the Americans were fighting to pacify the Philippines, the British were struggling to subdue the Dutch populations in Transvaal in the Second Boer War. The two vast Boer states combined had a sparsely distributed white population of just over 200,000 and were greatly outnumbered, militarily and demographically, by the British in the Cape colony to their south. Threatened with annexation into the British colony by force, the Boers initiated conventional preemptive attacks across their southern border and achieved initial successes. The British, however, rallied and reinforced, eventually attacking into the Transvaal against surprisingly stiff conventional resistance. The British poured in more and more troops, eventually reaching a theater force level of 479,000, and a year into the war managed to defeat the conventional Boer armies, only to see the Boers resort to widespread guerilla warfare. Unlike in America 125 years earlier, the British turned to scorched-earth tactics for the remaining year and a half of the war, destroying 30,000 Boer farms and imprisoning 120,000 Boers in concentration camps. About 8,800 Boer soldiers died in the war (4.4 percent of the population) and another 28,000 were taken prisoner, almost all shipped overseas to Saint Helena and other prisons. Death or overseas imprisonment was the fate of the majority of the Boer military-age male population. Boer civilian deaths, largely from malnutrition and disease, are estimated at 28,000, of which 22,000 were children under the age of sixteen.[1] Fully half of all Boer children would die during the war. This demographic catastrophe finally caused the Boers to accept English rule in 1902 and later incorporation into the Union of South Africa. Although resentment lingered for decades, there would be no third Boer war.

Tibet (1950–1959)

The sparsely populated, mountainous nation of Tibet had been, on and off, a province of China since the Mongol invasion in the thirteenth century. In modern times, Tibet once again became an independent nation in 1912, when Chinese troops left the country and the Dalai Lama returned from India to rule the country as both its spiritual and political leader. Tibet enjoyed self-rule until October 1950, when Mao reasserted China's historical claim and invaded Tibet with a force of 35,000 men. The 8,500-man Tibetan army was no match for the People's Liberation Army (PLA) and surrendered after suffering only 180 men killed or wounded. Pro-Tibet guerillas inflicted far greater casualties on the PLA, perhaps 10,000 dead from all causes. The war seemed over in May 1951, when the Dalai Lama and the Chinese agreed that Tibet would become an autonomously governed area of China. Several thousand Tibetans died in the eight-month war.

The peace would not last. Beginning in 1954, Khampa tribesmen of eastern Tibet rebelled against the Chinese in an all-out guerilla war that ended in thousands of deaths on both sides. In the Kangding Rebellion of 1956, as many as 2,000 Chinese troops were massacred by Tibetan separatists. The Chinese, for their part, bombed the monastery at Batang, killing 2,000 Buddhist monks and pilgrims. With CIA backing, revolt again broke out in 1959 when about 20,000 rebels took up arms in the capital, Lhasa, only to be brutally suppressed by the PLA. The rebellion ended with the Dalai Lama fleeing to India as the Chinese shelled his palace. Up to 87,000 Tibetans died in the 1959 uprising and the purges that followed.[2]

Chinese government statistics reveal that the Tibetan population declined 9.9 percent, from 2,775,000 to 2,551,000, between 1953 and 1964, the height of the rebellion and a period when the overall Chinese population was growing by nearly 20 percent.[3] True, thousands of Tibetans fled into exile, but by far the greatest causes of Tibetan population decline were the deaths due to the repressive counterinsurgency measures of the PLA and starvation among children. The Chinese census numbers clearly show that great numbers of military-age Tibetan males, born between 1915 and 1936, died during the decade of warfare in the 1950s, a result of military action, purges, and imprisonment.[4] The census data also shows very high childhood mortality for Tibetans born in the 1950s, a clear indicator of malnutrition and economic deprivation.[5]

As is the pattern in all other offensive wars, demographic catastrophe seems to be the proximate cause of the Tibetan collapse. Although Tibet has been comparatively tame since 1959, the 2008 demonstrations and riots attending the forty-ninth anniversary of the 1959 uprising show that Tibet's governance is still an open issue to many in Tibet and, indeed, throughout the world.

The Algerian War of Independence (1954–1962)

The Algerian war is a particularly interesting study for several reasons. First of all, it culminated a 130-year defense of Algerian sovereignty against French aggression that cycled intermittently through all the defensive strategies: conventional resistance, guerilla war, terrorism, civil disobedience, and passivity. Second, it shows the impermanence of an illiberal occupation based largely on force of arms. Finally, it demonstrates the depth of resistance of which Arab populations are capable.

France first invaded Algeria in 1830 as part of its colonial expansion. This first Algerian war lasted until 1847 and proved so costly to France that the French Foreign Legion was formed in 1831 as a way of deflecting the casualties away from the French citizenry. The restive Algerian Arabs were never completely subdued. Major pacification campaigns lasted through the 1850s, and there were major revolts against French rule in 1873 and 1881. French garrisons were numerous and busy. All told, the French army and the French Foreign Legion suffered perhaps a hundred thousand deaths, mainly from disease, in Algeria during the nineteenth century. Total Arab losses are unknown.[6]

By 1865 there had developed a fiction in France that Algeria was a part of metropolitan France, just another French province like Normandy. However, the indigenous Algerians, mainly Arabs, were second-class citizens in this new French land. The European immigrants, the *pied-noirs*, were the favored population for whom the Algerian colony was founded, and they disproportionately, if not exclusively, enjoyed the benefits of French rule.

As in Vietnam, World War II disrupted French colonial authority in Algeria, and the Algerians, many of whom had fought with the Allies against the Germans, hoped for a greater voice in the postwar era. Victory in Europe celebrations in Algeria on May 8, 1945, turned into the Setif Massacre, in which many of the jubilant Arabs grouped into mobs demanding self-rule and attacking the *pied-noirs*, killing about a hundred. The *pied-noirs*, the police, and the French military, in turn, retaliated by killing thousands, perhaps tens of thousands, of Arabs. Such was the pattern in Algeria. Nine million Algerian Arabs could not be controlled by 1 million *pied-noirs* without the most extreme means.

In 1954 the disaffection in the Arab community turned into a prolonged and escalating terrorist campaign and guerilla war aimed at the *pied-noirs* specifically and French rule in general. By 1959 the French military deployed almost 400,000 troops in Algeria, augmented by 26,000 gendarmes, perhaps 150,000 local Arab *harki* auxiliaries, and unofficial *pied-noir* militias. On the other side, the uniformed Algerian National Liberation Army (ALN) numbered perhaps 40,000 at any given time, most of whom were in Tunisia or Morocco training and organizing to infiltrate back into the country. Much of this force fought the French army in numerous battles and engagements in the border areas. The insurgency in the interior was mainly the work of the 30,000 regional troops and the 60,000-strong local militias that fought with or beside the ALN.

Over time, the French adopted harsh and successful counterguerilla and counterterrorist strategies, employing widespread torture of captives, reprisals against insurgent areas, and concentration and resettlement policies that displaced almost two million Algerians. The war reached its zenith in 1958, when the French were killing or capturing on average 3,500 ALN fighters per month. Typical French losses were a tenth or so of that figure.

By 1960 guerilla and terrorist attacks were in fact diminishing in the major Algerian population centers, if not in the countryside. But on the other hand, the war had seriously divided France, where the electorate simply no longer wished to commit its draftee sons to such a bloody, distasteful, and never-ending enterprise merely to preserve *pied-noir* privilege. Worse, the conflict had divided the French army, elements of which had become so wedded to the Algerian cause that they mutinied and developed secret armies of their own. The war descended into Frenchmen fighting other Frenchmen. In the end France sacrificed *pied-noir* interests to reach peace among themselves as much as with the Arabs.

When Algeria was granted independence in 1962, almost all of the *pied-noirs* and many of their *harki* supporters immediately fled the country. In terms of casualties, about 300,000 Arab Algerians, 141,000 of whom were ALN, died in the war. It must be remembered that this 3 percent death rate was felt mainly by that fraction of the Algerian population that actively supported the war for independence and was not evenly distributed across the general population. In insurgent areas, death and resettlement rates were extremely high. French and Foreign Legion deaths were about 25,000.[7]

The French war in Algeria is a case study in how even the most ardent counterinsurgent and counterterrorist strategies can be insufficient for resolving an insurgency in the absence of an agreed political formula for a new status quo. Except for the always problematic issue of the degree of coercion, indeed torture, that the counterterrorist French intelligence forces used to extract information from confirmed terrorists, the French army prosecuted the war in what was then considered, and still is, textbook fashion. Indeed, students at the U.S. Army Command and General Staff College today study the French approach for its lessons in counterinsurgency warfare. The French army deployed more than sufficient numbers of troops, raised numerous local auxiliaries for security and population control, effectively sealed the frontiers against ALN infiltration, conducted aggressive counterguerilla patrolling and raids against ALN militia in the interior, and effectively broke the National Liberation Front's (FLN) interior cellular terrorist and political structure, most notably in the celebrated Battle of Algiers.

Still this was insufficient in the long run. The French approach, in essence, was all stick and no carrot for the Arab majority in Algeria. For the overwhelming masses of Algerians, continued French rule meant only continued second-class status and *pied-noir* privilege, a formula no longer acceptable to a people inspired by post-World War II anticolonial success and renewed Arab nationalism. France had nothing to offer the Arabs but fear of French security forces, a policy certain to undermine the legitimacy of French colonial rule in both Algeria and metropolitan France.

The Chechen War (1994 to Present)

The Russians first invaded Muslim Chechnya in the eighteenth century and have been fighting to keep it ever since. The Chechens have always resisted Russian rule. Joseph Stalin resorted to forced deportations of Chechens and relocation of ethnic Russians into Chechnya, as well as other police state means, to keep the restive region in the fold. In the 1989 census, the approximately million people living within the borders of Chechnya numbered 715,000 ethnic Chechens and 269,000 Russians, the latter group mostly concentrated in the capital of Grozny.

As the Soviet Union dissolved in the early 1990s, the Chechens demanded self-rule, which they briefly obtained from 1991 through December 1999. The Russian campaign to reimpose rule from Moscow, which began in 1994 and still continues to this day, again illustrates how dogged a population can be in defending its independence against the most oppressive scorched-earth strategies. Again the Russians are the attacker, invading the first time in 1994, retreating in 1996, only to invade a second time in December 1999.

Although accurate casualty information is unavailable, Chechen deaths are generally reported to be more than 10 percent of the prewar Chechen population, perhaps far higher. The first Russian invasion caused at least eighty thousand Chechen deaths and the damaging or destruction of 30 percent of the country's housing. Perhaps an additional twenty thousand have died so far in the second invasion. Over a half million inhabitants of Chechnya have been made refugees. Most of the country was reduced to ruins. Chechen resistance, originally a conventional defense, then a guerilla war, continues now mostly as a terrorist campaign with no visible end.

As the guerilla war ebbed over the first years of the twenty-first century, Russia drew down the size of its army in Chechnya. Its goals were the complete withdrawal from counterinsurgency and counterterrorism commitments by the end of 2008 and the turning over of occupation responsibility to locally raised security forces.[8] In April 2009 the Kremlin announced "mission accomplished" and declared that the continuing "anti-terrorist" campaign had been successfully "Chechenized" under the leadership of the ethnic-Chechen strongman Ramzan Kadyrov, a Moscow appointee. Despite lifting the official "state of emergency," Kadyrov admitted that twenty thousand Russian Interior Ministry paramilitary troops would remain in Chechnya indefinitely.[9] This high ratio of security forces to population (two troops per one hundred population) suggests Moscow's continued need to control a restive population and ensure security for Chechnya's ethnic Russians and the Kremlin's Chechen allies.

Conclusions

There certainly is room for more research regarding foreign wars. Perhaps various cultures and nations have unique breaking points, both in terms of offense or defense, or are significantly different than the offered examples in their approaches to and tolerance for war fighting.

Such a hypothesis is only speculative, however, due to a lack of accurate data for most wars involving impoverished nations. Only wealthy nations can afford to accurately account for their populations. Surely there is some degree of self-selection in the offered examples; the struggles would not have become noteworthy had the defender simply conceded the issue of sovereignty. But there is nothing to suggest in any of the world's recent wars—whether between Iraq and Iran, or between Israelis and Arabs, or in the former Yugoslavia, or in Darfur or Somalia or elsewhere in Africa—that we have entered an era of "attrition-less" warfare. Not only is there overwhelming data supporting the assertion that offensive wars remain costly and prolonged, there is simply no body of contra-dictory evidence to suggest that modern offensive wars can be won cheaply or quickly. Nor is there any evidence that defenders are now more willing to yield sovereignty to invading armies than they were in the past. If Americans have become averse to the casualties of war, both friendly and enemy, we must realize that many of our potential enemies have not.

Chapter 6

The Balkan Wars: Learning
the Wrong Lessons

The wars in the former Yugoslavia that raged through the decade of the 1990s shaped the military's vision of post–Cold War conflict more than did any other single source. For most Americans, the Balkan Wars were a televised humanitarian tragedy, horrific for those involved but not a direct threat to America or its vital interests. Consequently, the American involvement was heavily diplomatic, with only episodic use of American military force at crucial times, force that always came in the form of air force and navy air strikes and missile attacks against Serbian targets. The army did not get into the fight, and contributed mainly by supporting United Nations and NATO peacekeeping forces with a small American contingent. The U.S. Army emerged from these conflicts as a declining equity in the Department of Defense portfolio.

From a doctrinal perspective the Balkan Wars illustrate the enormous difficulty inherent in imposing unwanted governance on resistant populations through offensive warfare. Serbia, led by the crude and brutal Slobodan Milosevic, was the aggressor in all of these wars. Slovenia, Croatia, Bosnia-Herzegovina, and finally Kosovo fought to defend their autonomous homegrown governments that had emerged after the death of Joseph Tito in 1980 and had strengthened in the anti-Communist, pro-freedom tidal wave that swept Eastern Europe in the late 1980s and early 1990s.

In many ways, the newly independent states of Yugoslavia were similar to Revolutionary America in 1775 or the Confederacy in 1861: citizens rallying to defend their local governance against a previously recognized authority that was now resorting to military means to impose stronger central government. As the American examples demonstrated, as do so many other examples around the world, citizens are capable of spontaneous uprising and defense against perceived tyranny.

When his initial quest to maintain Yugoslav central authority failed, Milosevic set his sights on creating a Greater Serbia, capturing the territories in Croatia and

Bosnia-Herzegovina where some 2 million ethnic Serbs lived either in their own enclaves or, more commonly, intermixed with ethnic Croats and Bosnians. Not wishing to include non-Serbian nationalities in his chauvinist utopia, Milosevic and his allies undertook the "ethnic cleansing" of non-Serbians from Serbian-populated regions of Croatia and Bosnia-Herzegovina. Soon, the non-Serbian parties to the war adopted similar "cleansing" strategies against the Serbs.

Later, Milosevic attempted to rid Kosovo of its ethnic Albanian majority. Serbs believe that Kosovo is the historical birthplace of the Serbian people. "Cleansed" of Albanians, it would have room to become a home for the hundreds of thousands of Serbians displaced by Milosevic's failed offensive wars.

Yugoslavia on the Eve of Civil War

In 1990, Yugoslavia was a nation of some 23.5 million people. Half of the population was Serbian, all but 2 million of which lived in Serbia proper with the remainder living mostly in Croatia (total 1990 population of 4.5 million), Bosnia-Herzegovina (4.4 million,) and Kosovo (2 million). Slovenia (2 million) was inhabited almost completely by Slovenes. Macedonia (2 million) and Montenegro (600,000) rounded out the unlikely federation of southern Slavs, formed from the ruins of the Austro-Hungarian Empire upon its demise in 1918.

Serbs, with the largest demographic share, had always been the glue holding the amalgam together. Since World War II, the strong dictatorial hand of Communist leader Joseph Tito suppressed whatever ethnic and separatist tensions boiled up. The death of Tito in 1980 and the weakening of the Soviet Union's grip on Eastern Europe during the 1980s led to greater regional autonomy in Yugoslavia as each of the Yugoslav republics developed popular local governments, complete with police and territorial defense forces.[1] In 1990, with Communist dictatorships irrevocably collapsing throughout Eastern Europe, all the Yugoslav republics held their own multiparty elections and installed popularly supported governments. It became clear during this outpouring of long-repressed sentiments that autonomy, if not outright independence, from Serbian rule had become the manifest desire of non-Serbians.

The two million ethnic Serbians living outside the borders of Serbia began to fear for their future. Protected up to now as the nation's predominant ethnic group, these Serbs recognized that dissolution of the nation would plunge them into minority status within the new states. The rejection of Serbian rule would likely find expression in anti-Serb activity.

Serbia elected Slobodan Milosevic as president in 1990 based on a political consensus that the federal republic must be maintained insofar as possible, the growing autonomy and separatism of the other regions must be halted and reversed, and the civil rights of Serbians everywhere must be protected. The tenor of the Milosevic bloc was much more pro-Serb than pro-Federation, and with naked brutality Milosevic and his henchmen would champion the Serb's ethnic claims while Yugoslavia crumbled around them.

The Ten-Day War in Slovenia

Both Slovenia and Croatia declared their independence on June 25, 1991. In a well-planned and -executed effort, Slovenian authorities employed their state police and territorial defense forces, perhaps 10,000 men total, to take over the border checkpoints with Austria and Italy, block the roads that the 150,000-man Yugoslavian National Army (JNA) would need to deploy its forces into Slovenia, and surround the few JNA units already stationed in Slovenia and keep them confined to barracks. Several German-supplied antitank and antiaircraft missiles provided the necessary firepower.

The JNA effort to prevent Slovene independence was halfhearted at best. Though commanded mainly by Serbs, the multiethnic rank and file of the JNA had no interest in fighting the Slovenes; many viewed the Slovenes as partners in the rebellion against Belgrade's authority. As the non-Serbs deserted their units, the JNA became paralyzed, more interested in surviving than attacking. Even the Serbs in the JNA questioned the need for war in Slovenia, where, unlike Croatia to the south, there were no ethnic Serbs to protect. So, at a cost of only eighteen Slovenes killed and having killed only forty-four of the JNA, Slovenia achieved its independence.[2]

Germany, based on historical and religious links with Slovenia and Croatia, immediately insisted that the European Community recognize the breakaway republics as sovereign countries.[3] Given the German and other European guarantees, Slovenia achieved the international recognition it needed and the Yugoslav federation was dead.

The Croatian War (1991–1995)

"Why should I be a minority in your state when you can be a minority in mine?" asked Yugoslav political theorist Vladimir Gligorov.[4] The JNA attack into Croatia was originally designed to keep Croatia in the Yugoslav federation, but it quickly became a more limited attack to liberate the Serbian populations in Croatia and bring them under Serbian military or paramilitary control. The Krajina, as it was called, consisted of the areas inside pre-1991 Croatia inhabited by substantial ethnic Serbian populations, namely the far eastern area along the Danube border with Serbia, including the city of Vukovar, and the much larger enclave in the south that ran along most of Croatia's border with Bosnia-Herzegovina. About six hundred thousand Serbs lived in the Krajina generally intermixed with equal numbers of Croats.

What remained of the JNA, supported by local Serb militias, attacked the Croats in the Krajina and made substantial, though often slow, gains. The conventional fighting between the JNA and the outgunned fifty-thousand-man Croat militia was sporadic, but ethnic cleansing and atrocity were constant. Resisting Croat cities were shelled to rubble by JNA artillery, and tens of thousands of Croats were killed or forced to flee.[5] Croat forces lost control of the Krajina in 1991 but were able to successfully defend the great majority, about 70 percent, of Croatian territory from Serbian occupation.

On January 3, 1992, Croatia and Serbia agreed to a cease-fire and Croatian independence, which was soon recognized by the United Nations and the European Union. Serbian Krajina, though a fact on the ground, received no international recognition. Milosevic, having achieved all he could for Serbian interests in Croatia, turned his attention to Bosnia-Herzegovina.

For Croatia, however, the loss of the Krajina and the Serbian brutality could not go unavenged. Over the next three years Croatia enlarged its army from seven to sixty-four brigades, which were armed primarily with surplus Soviet equipment purchased from former Warsaw Pact members. Many sources claim that retired senior U.S. military officers working for an American private security firm, Military Professional Resources International, using foreign donations from the United States and U.S.-aligned sources, greatly assisted the Croatian army in its rebuilding drive.[6]

In early August 1995, this new 150,000-man Croatian army launched Operation Storm, a surprise attack against the 40,000-man Serbian army in Krajina and routed it in four days. Through this offensive and the following negotiations, the Croatians recovered all of their lost territory. Most sources believe that at least 200,000 ethnic Serbs either fled into Serbia or were ethnically cleansed by the Croats. The International Criminal Tribunal for the former Yugoslavia has filed war crimes charges against both Serbs and Croats for their actions during the four years of the war. There has been no fighting between the two nations since 1995.

The War in Bosnia-Herzegovina

More than any of the other provinces of Yugoslavia, Bosnia-Herzegovina was the most ethnically diverse. In 1991 the Bosnia-Herzegovina population was 44 percent Muslim, 31 percent Serb, 17 percent Croat, and 8 percent mixed or other.[7] Worse, the ethnic groups were distributed in a patchwork of ethnic enclaves and in some areas were substantially intermixed. With no dominant ethnic majority and no apparent way to partition the country, Bosnia-Herzegovina provided the worst imaginable setting for a civil war, which from 1992 to 1995 became its fate.

On the heels of Slovenia's independence, Croatia's independence declaration and improving performance against the Serbian invasion, and Macedonia's bloodless secession, on October 15, 1991, the Bosnia-Herzegovina republican assembly declared the country a sovereign and independent state within its existing borders, rejecting calls from Belgrade for a smaller and redefined Yugoslav federation. Serbian delegates boycotted the vote that they considered illegal.[8]

Throughout the fall, all three ethnic groups in Bosnia-Herzegovina mobilized their military and paramilitary units. The Serbs, who had proclaimed the independence of four "Serb Autonomous Regions" within the borders of Bosnia-Herzegovina, requested and received great support from the JNA.[9] The Croatian minority in Bosnia-Herzegovina, having supported the October 15 declaration, began instead to fight for inclusion into the new Croatia.[10] For most of the war each of the three ethnic groups fought the other two. All factions conducted ethnic cleansing campaigns against enemy populations, with the Muslims generally getting the worst of it. In the end, the Croatians and Bosnians would reunite successfully against the encircled Serbs in the Krajina to help force a stop to the conflict in late 1995.

The proximate cause for the end of the war in Bosnia-Herzegovina was Deliberate Force, a fifteen-day NATO bombing campaign against Bosnian Serb military forces. The NATO planning for Deliberate Force began in September 1994, and in late 1994 NATO aircraft had actually fired at Serbian surface-to-air missile batteries in the Croatian Krajina as a demonstration of intent to intervene with air power should the Serbian ethnic cleansing in the region continue.[11] In July 1995 the Serb military overran the Bosnian areas of Srebrenica and Zepa and by August threatened UN peacekeepers protecting Gorazde. Renewed Serb artillery fire into Sarajevo was the last straw.[12]

On August 30, NATO aircraft based in Italy and flying from carriers in the Adriatic began striking Serbian military positions. Over the course of the next two weeks, NATO would launch more than 3,500 sorties into Bosnia, two-thirds of them American. More than a thousand bombs and cruise missiles were dropped or launched, 70 percent of which were "smart" munitions, inflicting substantial damage to the Bosnian Serb forces' antiaircraft units, ammunition supplies, bridges, communications, command and control centers, and military vehicle parks.[13] The bombing campaign ran parallel to diplomatic negotiations aimed at forcing the withdrawal of Serbian forces from specified areas where they threatened Bosnian populations. On September 14 the Serbs agreed to NATO's demands and began to move their forces out of Bosnia-Herzegovina. Satisfied with Serb compliance, NATO announced on September 20 that it would end the air campaign for the time being.

August and September 2005 were cruel months for Milosevic's dream of a Greater Serbia. The loss of the Krajina, the entry of NATO into the conflict, and increased diplomatic pressure convinced Milosevic that all that could be gained had been gained and that continued conflict would lead only to greater calamity. As a result, he agreed to negotiate an end to his wars in Bosnia-Herzegovina and Croatia. At the ensuing Dayton Conference (November 1–21, 1995) the leaders of the main warring factions—Serbian president Milosevic, Croatian president Franjo Tudjman, and Bosnian resident Alija Izetbegovic—pressured and assisted by American and international diplomats, hammered out the framework for the agreement that ended the war in Bosnia-Herzegovina. The agreement was essentially a map describing the various de facto ethnic zones and the establishment of a NATO peacekeeping force to maintain and extend the peace throughout Bosnia-Herzegovina.

Ironically, the numerous ethnic cleansing campaigns of the previous four years had, over time, consolidated and clarified the several Bosnia-Herzegovina ethnic zones, producing a mostly continuous, if imperfect, line of contact, along which a cease-fire could be based and UN and NATO peacekeeping troops could be deployed. The hope of Dayton was and is that with hostilities ended the citizens of Bosnia-Herzegovina would gradually learn to live together in peace again. The NATO peacekeepers, including American units, continue to enforce the peace between the various ethnic zones in Bosnia-Herzegovina to this time.

According to the U.S. Census Bureau's International Data Base, the population of Bosnia-Herzegovina declined by 700,000 people, a full 16 percent, during the years of the civil war. The confirmed death toll was at least 150,000, about 3.5 percent of the prewar population, and consisted mostly of killed Muslims and Croats but included 30,000 ethnic Serbs.[14] (Max Boot claims 300,000 deaths, about 7 percent.[15]) The other hundreds of thousands of lost souls were, presumably, either refugees or missing persons with unknown fates.

We can't tell from the data or events whether any of Bosnia-Herzegovina's ethnic groups had reached its demographic limit for self-defense by the time of the Dayton Accords. We do know that the parties to the civil war were sufficiently weakened by the conflict that no group attempted to restart the war after the withdrawal of the Serbian forces from Bosnia-Herzegovina.

Kosovo (1998–1999)

Serbians view Kosovo as the historic birthplace of the Serbian people, making it inseparable from Serbia proper. The 1991 census data, however, revealed a worrisome truth: many decades of adverse demographic trends had made Serbians only a small percentage of Kosovo's population, shrinking to only about 10 percent, while the more fertile ethnic Albanian population had grown to more than 80 percent.[16] Whatever Kosovo had once been, it was now overwhelmingly Albanian, and the remaining ethnic Serb minority protested that Albanian discrimination was forcing the few who remained to flee.

As early as 1987, Milosevic had begun championing the Kosovo Serb cause and promising them Serbia's protection.[17] Anticipating Kosovar separatist tendencies and wishing to protect the ethnic Serb minority, in March 1989 at Milosevic's behest the Serbian assembly abolished Kosovo's political autonomy, and in June

1990 Milosevic unilaterally abolished the Kosovo provincial assembly.[18] Belgrade, Milosevic determined, must rule Kosovo directly.

Kosovo's Albanians seethed, and there was some bloodshed, but Albanian political leaders largely capitulated to Serb demands and at least temporarily avoided the conflagration that engulfed Croatia and Bosnia-Herzegovina from 1991 to 1995. In response, separatist Albanians formed the underground Kosovo Liberation Army (KLA) in 1993. This organization remained unknown to most Kosovars until it burst onto the scene in 1998.[19]

Fanning the flames were extreme nationalists in Milosevic's government urging the restoration of the "proper" ethnic structure of Kosovo. According to the mythology of this very vocal element, Serbs had been the majority ethnic group in Kosovo prior to the Nazi invasion in 1941, when fascist antipartisan attacks against the Serbian population began to reduce the ethnic Serb faction. Conveniently linking two different enemies in a common cause, the Serbian nationalists further claimed that decades of illegal immigration from Albania, another state hostile to Serbia, only added to the problem that the Nazis had begun. The program of this extremist Serbian element called for expulsion of the "illegal" Albanian immigrants and their descendents and resettlement into Kosovo of the six hundred thousand Serb refugees from the wars in Croatia and Bosnia-Herzegovina, then residing in temporary accommodations within Serbia proper. In this manner, the "population ratio that existed prior to April 6, 1941" could be restored.[20] Indeed, sixteen thousand Serb refugees had already been resettled in Kosovo, and these *Shiptars* became both a source of Albanian fears and a target of Albanian wrath.[21]

The KLA initiated terrorist activities in the spring of 1998, targeting Serbian police, politicians, and symbols of Serbian rule. Serbian reprisals were immediate and severe, and in very short order Kosovo polarized into irreconcilable camps. The KLA, like any good guerilla organization in its early phase, was everywhere and nowhere, fighting when it could, melting away when the situation demanded. The Serbs, for their part, were incapable of moderation, and by late summer two hundred thousand Kosovars (10 percent), mostly ethnic Albanians, had been displaced from their homes, with a quarter of them living outdoors in squalid camps and about half of them fleeing the province.[22] NATO, not wishing a replay of the Bosnia-Herzegovina disaster and clearly remembering how air strikes had forced Milosevic to buckle in 1995, prepared for and threatened air strikes to stop this latest Serb ethnic cleansing campaign.

Though no bombs fell, on October 12 Milosevic agreed to end his offensive and withdraw Serb forces in Kosovo down to prewar levels. The Serbs largely complied with the agreement, and many of the refugees returned to their homes. With them came the KLA, which asserted more control than ever and prepared for the next round.[23] The terrorism and counterterrorism continued through the winter, albeit at a reduced level. By March 1999 at least two thousand Kosovars had died in the fighting.

Also over the winter, the Bill Clinton administration and NATO, based on the Dayton precedent, insisted that the Serbs and Kosovars meet to negotiate a permanent governance formula for Kosovo, to be enforced by NATO peacekeepers. Both parties resisted, the Kosovars not desiring any status short of complete independence and Milosevic unwilling to backtrack on the unitary nature of Serbia and Kosovo. Ultimately, the Kosovars reluctantly agreed to sign the Rambouillet Agreement, giving them broad autonomy, but Milosevic did not.[24] NATO threatened imminent bombing, but Milosevic didn't budge. NATO calculated that a day or two of light air strikes would change his mind.[25] Milosevic calculated differently. His plan, Operation Horseshoe, was to use Serbian forces to empty Kosovo of Albanians "within a week," changing the facts on the ground and rendering the air attacks irrelevant.[26] Both sides overestimated their positions.

Operation Horseshoe began on March 1, 1999, the Serbs attacking with 40,000 troops, including tanks and artillery, into the central Drenika region of Kosovo, destroying KLA units and driving tens of thousands of Albanians from their homes.[27] The NATO air campaign began on March 24 and throughout its seventy-eight days was playing catch-up with the ever-worsening calamity on the ground. By the time the war ended, almost 850,000 Kosovars had fled or been deported to other countries, and hundreds of thousands more were internally displaced within Kosovo.[28]

The figures suggest that perhaps as many as two-thirds of Kosovo's Albanians had been driven from their homes in the Serb offensive. The plight of the refugees only steeled NATO's resolve. The number of aircraft assigned to the air campaign rose from less than four hundred to well over a thousand as the weeks wore on; daily sorties grew from three hundred to, at one point, nearly nine hundred.[29]

More importantly, the targets shifted from an initial concentration on air defense and other military targets to the destruction of the economic infrastructure of Serbia herself. By mid to late April, NATO aircraft and cruise missiles were destroying oil refineries, petroleum depots, the Danube bridges, railway lines, radio and television stations, and automobile factories. Later, Serbia's electricity plants would be added to the target list. One Serbian economist confessed that the results of the bombing were an "economic catastrophe" and predicted that Serbia's "industrial base will be destroyed and the size of our economy cut in half." [30]

NATO's frustration, though, was its perceived inability to stop the ethnic cleansing campaign in Kosovo, no matter how much it punished Serbia. Without a way to defeat Serbian army units operating in Kosovo, the destruction of Serbia might be pointless. A ground force was required in Kosovo, and within days of the start of the bombing campaign NATO commander Wesley Clark asked the U.S. Army to provide a contingent of attack helicopters to help hold back the Serbian onrush.[31]

The army's response was Task Force (TF) Hawk, an impromptu assortment of tank-killing Apache attack helicopters, Abrams tanks, Bradley infantry fighting vehicles, artillery, engineers, air defense, and logistics support. Unable to deploy the TF Hawk directly into the Kosovo battle area, this unwieldy 5,500-man U.S. force was assembled in Albania, far from the fighting in Kosovo. In the end, the air force flew five hundred C-17 sorties to deploy the task force from Germany to the very poor airfields in Albania. By mid-May, with the war half over, TF Hawk was at last ready to fly attack helicopters into Kosovo in support of the desperate Albanians, but the chairman of the Joint Chiefs of Staff, Hugh Shelton, judged the risks too high.[32]

Avoiding NATO losses was a key component of the NATO plan. While fixed-wing fighters and bombers using precision guided munitions could fly high to avoid the Serbian air defenses, Apaches had to fly low, where they were vulnerable. Without friendly forces or artillery fire to secure their battle positions, the Apache pilots would be sitting ducks, and losses could be high.

So for the remainder of the war, the only ground force in Kosovo opposing the Serb offensive was the outgunned and outnumbered KLA. To be sure, the KLA benefited from NATO air attacks against the Serbian army in Kosovo, which kept the Serb's mechanized forces off the roads and dispersed for protection, reducing their effectiveness. There were also several instances when NATO pilots actually discussed their attacks with KLA forces on the ground. But for the most part, the NATO tactical air strikes lacked the necessary air-ground coordination, namely forward air controllers, to be truly effective, and the KLA could never put enough pressure on the Serb army to force it to mass and maneuver, which would have made Serb units obvious targets from the air.[33]

In the end, on June 9, after suffering seventy-eight days of bombing, having been on the receiving end of some thirty-three thousand air sorties and over five hundred cruise missile strikes, Milosevic consented to withdraw his forces from Kosovo, allow a 50,000-man NATO peacekeeping force to occupy Kosovo, and accept the governance formula in the Rambouillet Agreement. Factors that weighed into his decision included the realization that NATO could bomb Serbia indefinitely; pressure from the Serbian oligarchy, who knew that continued bombing would bring them bankruptcy[34]; falling morale and rising desertions among Serbian draftees in Kosovo[35]; the manifest resolve of NATO leaders to invade Kosovo, indeed Serbia herself, with 150,000 to 300,000 troops if Milosevic didn't capitulate from the effect of the air campaign alone[36]; and finally his abandonment by Russia, his only ally, which could provide no meaningful assistance.[37]

For all their bluster about creating a Greater Serbia, the Serbian people were now staring into the abyss of defeat and economic catastrophe. Now a pariah state led by an indicted war criminal, Serbia escaped with little more than its sovereignty intact. The only silver lining was that the country's costly eight-year struggle to save Yugoslavia was now over. For his efforts, Milosevic was voted out of office in 2000 and in 2001 was turned over to UN authorities to stand trial for war crimes.

Measured from the beginning of Operation Horseshoe, the toll of the Kosovo war was about 10,000 ethnic Albanians dead and missing. The Milosevic government reported only 1,800 casualties, including about 750 soldiers and policemen killed. NATO estimates of Serb casualties were far higher: 5,000 killed and 10,000 wounded.[38]

In retrospect, Serbia's wars against its neighboring republics were all offensive wars of occupation. Serbia was never attacked by its neighbors. Never was Belgrade's rule over the ten million Serbs within the borders questioned or threatened, except perhaps by NATO at the very end. For all the existential rhetoric, insofar as Serbia was concerned the wars had limited objectives: maintaining the federal union, protecting Serb minorities in adjacent republics, expanding its boundaries to absorb ethnic Serb enclaves, and maintaining its historical shrines in Kosovo. In all these wars Serbia demonstrated a calculus for knowing when to quit, knowing when the cost of its ambitions had exceeded the potential rewards. Reflecting the lack of vital Serbian interest, draftees from Serbia performed halfheartedly in Croatia and Kosovo, often deserting or shirking from combat, causing the 1991 and 1999 offensives to fail to reach their objectives. Indeed the local Serbian militias, fighting a true life-and-death struggle, did the heaviest fighting and the real dirty work in the ethnic cleansing campaigns in Croatia, Bosnia-Herzegovina, and Kosovo.

When challenged, Serbia quit. Slovenia achieved its independence without significant fighting, as did Macedonia. Croatia cleared the Krajina in its surprise 1995 offensive, and Serbia did nothing. Two weeks of NATO bombing caused Milosevic to stop his campaign in Bosnia-Herzegovina. Much greater bombing was required to stop the 1999 invasion of Kosovo, but still Serbia conceded defeat rather than risk further economic destruction or NATO invasion. In the totality of his actions, Milosevic seemed to pursue his aggression only as long as the cost was low, retreating when the cost to Serbia became too high. Conversely, the attacked republics succeeded in their defensive wars by inflicting casualties and destruction on the Serb forces or, when they lacked the military capability, getting NATO to provide the needed firepower. Raising the cost to the attacker beyond his threshold of tolerance was, as ever, the surest path to defensive success. In most respects the wars in the former Yugoslavia conform to the discussion of offensive and defensive war in chapter 2. They were not a new form of warfare, just a return to the forms of warfare that had plagued Europe prior to the Cold War.

Lessons Learned and Mislearned

Militaries worldwide were impressed—dazzled, really—by the effectiveness of the NATO air campaign in the Kosovo war, especially the tremendous effectiveness and efficiency of America's precision-guided munitions: the cruise missiles and smart bombs. Although these munitions did not achieve the "one bomb per target" ideal, they did achieve a remarkable "less than two bombs per target" level of efficiency. Targets, such as bridges and factories, that once would have been serviced by squadrons of aircraft dropping hundreds of dumb bombs, were now invariably destroyed by only a sortie or two, and by historical standards collateral damage was minimal.

The U.S. Air Force, the undisputed star of the war, could hardly contain its delight. The success of the air war during Desert Storm in 1991, the success of the two-week air offensive in stopping the war in Bosnia-Herzegovina in 1995, and the capitulation of Serbia to the much larger air campaign in the Kosovo war were seen as proof that the air arm could win wars with or without the support of the army. Newspaper editorials across the nation acclaimed the air force. The *Los Angeles Times* wrote that NATO had achieved "what some military experts had predicted was impossible; a victory achieved with air power alone."[39] The *Wall Street Journal* concurred that "air power alone can win some kind of victory."[40] The *New York Times* called the air campaign "a success and more—a refutation of the common wisdom that air power alone cannot make a despot back down."[41] Former army officer turned defense analyst Andrew Krepenivich joined the chorus, "[A]lmost alone, American air power broke the back of the Yugoslav military and forced Slobodan Milosevic to yield to NATO's demands. What air power accomplished in Operation Allied Force would have been inconceivable to most military experts 15 years ago."[42]

Although the air force would not go as far as the press did in dismissing the army contribution and did admit that its tactical attacks against the Serbian Third Army in Kosovo were frustrated due to lack of an effective ground force, the air force still felt that it alone had won the war, and could have won it faster had the army's General Clark not focused the initial air effort against the Serbian armies in the field. Lieutenant General Michael Short, the commander of NATO air forces during the Kosovo war, later offered the following Clausewitzian analysis:

> The Third Army in Kosovo was not the center of gravity. . . . I would have defined the center of gravity . . . as Slobodan Milosevic . . . and the infrastructure that kept him in power. We need[ed] to pressure him and

the people around him and that infrastructure around him and make it very, very painful for them to continue the conflict. . . . I continue to believe that we could have avoided the war of attrition if we had been allowed to attack the center of gravity from the first strike and had the courage to stay the course. . . . If you want to guess, I'm saying less than 30 days if we were allowed to go where we need[ed] to go. The first night the power grid goes down and the bridges go down, the command and control nodes go down . . . and the machine kicks in—"Slobo, this was just the first night." . . . [Of course, it] is impossible to defeat an enemy from the air alone if he is willing to absorb any amount of punishment and is willing to allow his country to be destroyed rather than accept terms of surrender. [43]

The implication was clear: armies are meaningless; air power alone can destroy nations and compel surrender.

The U.S. Air Force was not alone in drawing this conclusion. Some Russian military observers also concluded that precision-guided munitions had changed the nature of warfare. Major General Vladimir Slipchenko of the Russian Academy of Military Sciences called precision-guided munitions, especially the cruise missile, the sixth generation of warfare, as transformational to the way wars are fought as were the inventions of metal weapons, gunpowder, machines, and nuclear weapons. According to Slipchenko, in the 1990s the Americans had demonstrated the ability to wage "non-contact, long-range, precision warfare," which could devastate countries and win wars without the ground forces of the two sides ever battling each other along a line of contact on the ground. Slipchenko assessed the lessons of Kosovo as follows:

High precision cruise missiles were launched from a distance of 80-800 kilometers and struck all targets very accurately. There were 900 economic and military infrastructure targets in Serbia and Kosovo, and the 1,500 high precision cruise missiles sent there were something like 75-80% effective against them all . . . With respect to the regime in Yugoslavia, the people themselves changed it . . . What did the Americans do: they destroyed 80% of Yugoslavia's economic potential and the Yugoslav people themselves changed the political regime. The armed forces were unscathed but they fell apart and now are virtually nonexistent because there is no economy. That is remote war victory for you. To win a contact war would have required, first of all, defeating the opponent's armed forces, then destroying its economic potential, and then changing the regime. Now that's all done the other way.[44]

* * *

106

Slipchenko's assessment was that fourth-generation mechanized ground forces were not important in the new era of sixth-generation noncontact warfare and recommended that the Russian military transfer the ground forces to the "border guards" to "repulse aggression in the border region" or to the Interior Ministry for population control within Russia herself.[45]

Similarly, the U.S. Army struggled for relevance after the embarrassment of Task Force Hawk, and there was a growing perception within the army that it had to do something quickly to get a player in the game. As new army chief of staff Gen. Eric Shinseki put it in June 1999,

> Our heavy forces are too heavy and our light forces lack staying power. Heavy forces must be more strategically deployable and more agile with a smaller logistical footprint, and the light forces must be more lethal, survivable, and tactically mobile. Achieving this paradigm will require innovative thinking about structure, modernization efforts, and spending.[46]

Fixated on solving the Task Force Hawk problem, the army concentrated its efforts along two mutually supporting lines of effort.

The first was to design a brigade that could deploy and fight independent of a cumbersome divisional base. By October 1999 Shinseki ordered the creation of two experimental brigades, later known as the Interim Brigade Combat Teams, at Fort Lewis, Washington. By November 2000 the army had decided to equip these brigades with the wheeled Stryker combat vehicle, based on the eighteen-ton Canadian light armored vehicle (LAV III), and signed contracts that would deliver Strykers to the army beginning in spring of 2002.[47]

In the meantime, the army was working toward its perceived definitive solution, the Future Combat System (FCS), a high-tech, lightweight combat vehicle that would be the army's primary weapons platform for the future. By May 2000 four separate defense industry teams were competing for the FCS contract and hoped to begin producing the new vehicles in 2008.[48] At over $160 billion, the FCS program would be the most expensive weapons procurement in army history. However, neither of these two combat vehicle programs would make a marked difference in the imminent global war on terror.

Even though Shinseki was expending great energy to get momentum behind his transformation of the army, his efforts to make it more agile in dealing with small-scale conflicts such as the Balkans were too little and too late, not to mention beside the point for the new Republican administration elected in November 2000. Having run a platform belittling Clinton-era "nation-building" and promising a leaner but tougher military, the newly inaugurated president, George W. Bush, immediately fired a warning shot across the army's bow. "We are witnessing a revolution in the technology of war," said Bush. "Power is increasingly defined not by size but by mobility and swiftness. . . . [S]afety is gained in stealth and force is projected on the long arc of precision-guided weapons. . . . [U.S. air power] will be able to strike across the world with pinpoint accuracy, using both aircraft and unmanned system. . . . [The Navy] would connect information and weapons in new ways, maximizing our ability to project power over land."[49]

Though the president praised Shinseki's fledgling transformation efforts using words similar to those in the army's public relations publications, it was clear that the army was of little immediate use in the new administration's vision of defense. Bush ordered Secretary of Defense Donald Rumsfeld to conduct a "top to bottom" review of defense policy, strategy, missions, and spending and provided Rumsfeld "a broad mandate to challenge the status quo."

In the months immediately preceding the 9/11 attacks and the subsequent decision to invade Iraq, the U.S. Army was literally on the chopping block. As the Quadrennial Defense Review proceeded into the late summer of 2001, the leading proposal coming from the defense secretary was the elimination of two of the ten army divisions then on active duty.[50] Having painfully drawn down in the 1990s from eighteen divisions (780,000 soldiers) to ten divisions (485,000 soldiers) after the conclusion of the Cold War and Desert Storm, the army staff spent most of 2001 trying to explain to the Department of Defense why the army should not be reduced by 50,000 or so additional soldiers and struggling to come up with ways to accomplish the reduction should it be ordered. The army's contention that "it's pretty difficult to bring effects to bear from standoff ranges when the enemy is using sanctuary . . . . you need boots on the ground"[51] rang untrue in the aftermath of the Kosovo war. Besides, for an administration looking for deep cuts in what were deemed unneeded military programs to pay for other priorities, defense and nondefense, the army's argument that it needed every single pair of those 485,000 boots was just too much to swallow. In the new age of "noncontact warfare," the army was slow to realize that it was now playing only a bit part in the nation's defense against potential aggressors.

Worse still, after 9/11 Rumsfeld failed to grasp that the army would have to become the decisive force in our new offensive wars to occupy and remake Afghanistan and Iraq. Rumsfeld errantly believed that these conflicts would test and vindicate his theories of rapid and decisive warfare and would require the smallest imaginable army footprint. Even with America now at war, Rumsfeld would not consent to increasing the size of the army. Thinking that the defensive striking power that had turned the tide in the Balkans would be sufficient to win the offensive wars in Afghanistan and Iraq, Rumsfeld failed to reorganize the Department of Defense for offensive war and occupation duty.

Of course, neither the joint staff nor the army had relevant doctrine disputing Rumsfeld's notions. Indeed, the air force and navy largely agreed with Rumsfeld's assessments, which favored them bureaucratically, while the army was working on its own Clausewitzian concepts of rapid, decisive warfare and was institutionally against becoming bogged down in nation-building missions, such as occupation. These flawed ideas, and institutional biases that sustained them, would become painfully exposed during the course of our two upcoming invasions.

Chapter 7

Iraq: A Case Study in Postmodern Military Failure

> Discovery commences with awareness of anomaly, i.e., with the recognition that nature has somehow violated the paradigm-induced expectations that govern normal science. It closes only when the paradigm theory has been adjusted so that the anomalous has become the expected.
>
> —*Thomas S. Kuhn*

We hoped in Iraq to demonstrate the power of America's post-Cold War military. Instead, Iraq exposed the narrow-mindedness of the contemporary American professional military and its prevailing doctrines. The inability of the world's strongest and most lavishly resourced armed forces to produce a permanent favorable outcome in Iraq despite over a decade and a half of effort, hundreds of billions of dollars, thousands of American lives lost, and hundreds of thousands of Iraqi deaths certainly suggests that the army should revisit its dominant paradigms for relevancy and suitability in the contemporary environment. In an even larger sense, the war in Iraq challenges traditional American notions of "war" and "victory," substituting instead an open-ended condition of "persistent conflict."[1] The final scenes in Iraq are yet to be written, but the tremendous battle successes in the opening acts have not yet translated into a better world for either Iraqis or Americans. American hubris has morphed into discouragement, not just once but twice.

Desert Storm

The 1990–1991 Gulf War, known as Desert Shield/Desert Storm, was at first a defensive war to protect Saudi Arabia, then a counteroffensive to liberate Kuwait from Iraqi aggression. The American Cold War armed forces, designed to defeat a potential Soviet assault against NATO in Europe, instead found themselves tasked to defeat Saddam Hussein's large Third-World military, which improbably had invaded Kuwait and threatened the vital Saudi oil fields. President George H. W. Bush assembled a vast United Nations-sponsored coalition to reverse the Iraqi fait accompli. At its peak, U.S. Central Command (CENTCOM) commanded 541,000 American service personnel in theater and exercised at least partial control over perhaps 100,000 additional coalition forces from our various allies and friends worldwide. Iraq opposed CENTCOM with forty-two divisions, which on paper comprised 350,000 soldiers, though only about 200,000 Iraqis were present for duty with their units when the actual war began.

The complete qualitative overmatch that the American and NATO forces enjoyed was evident from the opening ground skirmishes and the forty-day preliminary air campaign. The coalition's hundred-hour ground assault resulted in the nearly complete destruction of the Iraqi military deployed in the Kuwaiti theater. Estimates of Iraqi military dead range from 40,000 to 100,000 against only 146 American battle deaths from all services, a quarter of which were from friendly, not Iraqi, fire. Never in history had a battle been so lopsided.

General Norman Schwarzkopf, the CENTCOM commander in chief, was an adherent to the Clausewitzian paradigm that pervaded the army's doctrine in the 1980s. Schwarzkopf identified Saddam Hussein himself and the Republican Guard as the two enemy centers of gravity and designed his plan around isolating the dictator and destroying the Republican Guard divisions through fire and maneuver.[2] At his victory's-eve February 27, 1991, press conference from his headquarters in Riyadh, Schwarzkopf dismissed Saddam as being "[un]schooled in the *operational art*," the army's shorthand for the synthesis of Clausewitz's ideas that were then in vogue. Brilliant though it may have been, Schwarzkopf's battle of annihilation over the hapless army of an Arab dictator did not produce an acceptable postwar peace.

Indeed, there is little indication that the American military establishment in the Middle East or Washington had clearly thought through the problem that postwar Iraq would present. As President Bush and his national security advisor, Brent Scowcroft, later remarked, "The end of effective Iraqi resistance came with a rapidity which surprised us all, and we were perhaps psychologically unprepared for the sudden transition from fighting to peacemaking."[3]

Schwarzkopf defends himself by observing, truthfully, that the UN mandate called only for the liberation of Kuwait, not regime change in Baghdad. Still, even given the limited nature of the war, the Americans had a responsibility to ensure a peace commensurate with the victory, and this was not done. Schwarzkopf consented to Bush's idea of stopping the ground war at exactly one hundred hours—a public relations gimmick!—resulting in the escape of many of the Republican Guard forces in theater.[4] Astonishingly, Schwarzkopf and Washington had not meaningfully discussed or agreed on the cease-fire terms that would be imposed on the Iraqi government and military. A *Frontline* interview years later revealed the lack of coordination:

> Q: Safwan. Did you get any guidance from Washington as to how you were going to conduct [the cease-fire discussions at] Safwan?
>
> SCHWARZKOPF: Fundamentally, no. Colin [Powell] and I talked about the objectives that we thought we wanted to accomplish there. Because I didn't have any terms of reference I wrote my own. I sent them back to Washington and basically got them back unchanged saying "Go to Safwan and do that!"
>
> So I went to Safwan and the President announced that we were going to discuss the terms of the ceasefire and so therefore, I went to Safwan with my own instructions which, basically, number one was to get our POWs back, and then number two to make sure that we had very very clear lines drawn so that we didn't have any inadvertent battles after that.
>
> Q: White House gossip says, "You know, we tried to offer Norman Schwarzkopf lots of advice about Safwan but he wouldn't hear it. He wanted his own ceasefire."
>
> SCHWARZKOPF: That's absolutely bogus. I mean, anyone who would say that they offered Norman Schwarzkopf anything going into Safwan, that's a blatant lie.[5]

Complicating the postwar picture was President George H. W. Bush's call for revolt within Iraq, most notoriously his February 15, 1991, plea for the "Iraqi people to take matters in their own hands to force Saddam Hussein, the dictator, to step aside," a message beamed by the *Voice of America* into Iraq. The Kurds in the north and Shiites in the south did revolt, as the American president had urged, only to be driven from their homes and brutally suppressed by Saddam's forces, many of whom had escaped back into Iraq courtesy of Bush's hundred-hour cease-fire decision and Schwarzkopf's hip shots at the cease-fire discussions at Safwan. Again from the *Frontline* interview:

> Q: What was your advice as to what to do about the uprising, the Shi'ite uprising?
>
> SCHWARZKOPF: Well I don't know that I was asked for any advice. And as a matter of fact, I'm quite sure I wasn't asked any advice. . . .[6]

In Washington and CENTCOM, the focus on winning the force-on-force battle had crowded out timely discussion of crucial postwar policy, leading to ad lib decisions, confusion, and recrimination at war's end. Ironically, Schwarzkopf's cease-fire meeting with the Iraqis at Safwan is the pictured example of conflict termination in our current joint doctrine.[7]

While the American forces deployed out of theater in 1991, Saddam waged a vicious war against the rebellious Kurds and Shiites, killing an estimated twenty thousand to thirty thousand of them while Bush backtracked from his reckless remarks. Finally, as Saddam's forces drove Iraq's Kurds—homeless, starving, and freezing—into the mountainous fringe of northern Iraq, Bush finally intervened with Operation Provide Comfort to prevent further genocide, and in so doing unwittingly became the sponsor of an autonomous Kurdish region within Iraq. Worse for the Bush administration, Saddam remained in power, and no amount of UN arms inspectors, no-fly zone enforcement actions, retaliatory cruise-missile strikes, or oil export limitations could depose him.

The festering dissatisfaction over the results of Desert Storm over time bore the seeds of our successor war in Iraq, Operation Iraqi Freedom. Throughout the remainder of the 1990s, there was a growing consensus among the emerging neoconservative wing of the Republican Party, and indeed many Democrats, that regime change in Iraq must be put back on the table. In October 1998 the Republican-led Congress passed and President Bill Clinton signed into law the Iraq Liberation Act, which pledged American support to groups seeking to overthrow the Hussein government. In the wake of 9/11 it was this neoconservative consensus that opportunistically steered the war against al-Qaida into a war to depose Saddam Hussein, even though there was no apparent threat from Iraq at that time. Elevated in power during the new administration of President George W. Bush, the Republican hawks, most notably Vice President Dick Cheney (the Desert Storm secretary of defense) and Deputy Secretary of Defense Paul Wolfowitz (the Desert Storm assistant secretary of defense for policy), would finally get another opportunity to achieve the complete victory denied them ten years before.

Operation Iraqi Freedom: Victory Yields Chaos
Unlike Desert Storm, Operation Iraqi Freedom (OIF) would not be a defensive war, a counteroffensive battle to liberate Kuwait, but an offensive war waged to bring about new governance in Iraq. The conventional war, from the crossing of the Kuwait border on March 19, 2003, to the fall of Saddam Hussein's statue in Baghdad's Firdos Square on April 9, would last only twenty-one days.

The "march up-country" was a workmanlike deliberate attack conducted by the smallest practical ground force into a complex and confusing Iraqi defense in depth.[8] With great competence and minimal casualties, the army and marine forces fought their way through sporadic Iraqi resistance up the only two available approaches to Baghdad, averaging over a kilometer an hour in their advance. Casualties in the initial occupation of Iraq were light; about 139 Americans died in OIF during the first six weeks. Any military professional would assess coalition performance in the maneuver part of the war to be about as good as it gets. Reflecting the general consensus both in the country and in the U.S. Army that the hard work was done, on May 1 President Bush flew to the carrier USS *Abraham Lincoln* and under a banner that read "Mission Accomplished" declared the end of major combat operations in Iraq.

The week before, CENTCOM commander in chief Tommy Franks, viewing *his* mission as over, had decided to retire from the army.[9] Army combat units in Iraq were getting their orders to return to Kuwait to prepare to go home. Some 3rd Infantry Division soldiers who led the march to Baghdad were told by their chain of command that the war was over and they would be back in the States in sixty days.

Only in the most narrow military sense could the war have been seen as over and won, and it became clear as the months wore on and the chaos in Iraq grew that the march to Baghdad and the deposing of Hussein's regime was only the preliminary round to what would be a much greater and more costly effort: imposing new governance on the Iraqi people, a task that the military did not even recognize as its responsibility but, as of this writing, has ended up costing the army and marines over four thousand additional lives and, in the assessment of some experts, is stretching the army to the breaking point.

In a way, the failure of the American military establishment to adequately consider the requirements for rebuilding postwar Iraq was the perfect storm, the merger of the administration's desire for a short, cheap war with the military's institutional preference to focus solely on war fighting and avoid nation-building exercises. Mutually reinforcing their respective wishful thinking, the civilian leadership and uniformed military virtually ignored their inherent responsibility under international law of occupying armies to provide governance in the wake of their invasions. Larry DiRita, Rumsfeld's spokesman, told postwar planners, "We don't owe the people of Iraq anything. We're giving them their freedom. That's enough. . . . We're going to stand up an interim Iraqi government, hand over power to them, and get out of there in three to four months. All but twenty-five thousand soldiers will be out by the beginning of September."[10]

The Bush candidacy had railed against the Clinton-era nation-building operations in the Balkans and Somalia and was determined not to repeat those errors in Iraq. In the neoconservative mythology, we were "liberating" Iraq, not "occupying" Iraq, so reestablishing governance could be handled by the locals. Franks explained his April 2003 thoughts as follows: "Would Iraq make it? I didn't know. But I knew that now, because of what our troops had done, almost twenty-six million Iraqis had a *chance* [original emphasis] to build a new nation."[11]

One would have expected a greater commitment to success, a recognition that building a better Iraq had become an American national obligation that demanded his most diligent efforts. But Franks, like so many other officers of his military generation, was a Clausewitzian, who, from the first discussions of an invasion of Iraq in early December 2001, viewed the upcoming battle in terms of applying "the necessary lines of operation to attack or influence what Clausewitz has described more than a hundred years before as the enemy's 'centers of gravity.' "[12]

For Franks, the focus was always on defeating the enemy, not on the governance that followed.

All of this was reflected in the insufficient planning and resources devoted to the occupation. Planning for the post-conflict, termed Phase IV, was delegated to ad hoc planning groups, but they lacked the needed support of the civilian cabinet or military commanders and could find no current military doctrine concerning occupation. These uncoordinated groups produced thousands of pages of documents and PowerPoint slides but no executable plan or enabling force structure.[13]

Only very late in the game did the administration recognize that Phase IV was at risk. Less than two months before the 3rd Infantry Division crossed the berm to head to Baghdad, Jay Garner, a retired lieutenant general who had served in Operation Provide Comfort in the early 1990s, was enlisted to create the Office of Reconstruction and Humanitarian Affairs (ORHA) to begin the rebuilding of Iraq. As he relates it, he was sitting in a restaurant in New York City toward the end of January when:

> I got a call from Rumsfeld's office saying, "We'd like to talk to you. We want to talk to you about doing something in postwar Iraq." So I went to see them. . . . Rumsfeld said, "Tommy Franks and I really want you to do this". . . . I went to work for Rumsfeld around 27 January and spent from 1 February to 15 March in the Pentagon. . . . By 16 March we had close to 300 people. We'd grown from one person to 300 people in about six weeks, and we deployed to Kuwait. From 16 March to 21 April, we stayed in Kuwait. . . . On 17 April I flew to Doha to see Tommy Franks, and I said, "You've got to get me and my team into Baghdad, into Iraq." . . . Now, Tommy Frank's plan was that we would go into Phase IV in about anywhere from 30 to 90 days after combat operations, then we would take our whole team and put it in there. I said, "You know, that plan doesn't work anymore." . . . On the 18th he called me back. He said, "Jay, you're free to go. God bless you." . . . Then we road marched into Baghdad.[14]

By the time Garner's ORHA convoy of several hundred Chevy Suburbans reached Baghdad on April 23, fourteen days after the fall of Saddam's statue, the city had already suffered through two full weeks of looting and chaos, and the governmental apparatus and infrastructure of the Iraqi capital were in ruins.[15]

Even though the Americans had occupied the capital and the Iraqi government had dissolved and fled, no American had declared martial law. There had been no proclamations of occupation posted for the Iraqis to read, no broadcasts of ordinances that the Iraqis should follow, no military governance teams to organize local governments and police forces. Having disregarded the American pattern of occupation that began with Maj. Gen. Winfield Scott's campaign in Mexico and reached maturity in World War II, the U.S. Army naively assumed that their contribution to the war effort—winning battles—was over. American troops, spread dangerously thin, were unprepared for the lawlessness that followed the collapse of the Saddam regime and were unclear of their authority to stop the looting, or even the necessity of doing so. Most just stood and watched. On April 16, after a full week of anarchy, General Franks (still operating from his headquarters in Qatar) issued his little noticed "Freedom Memo," his initial greetings and instructions to the Iraqi people. It had no discernable impact.

General David McKiernan, the CENTCOM land component commander, experienced the mayhem firsthand when he finally arrived in Baghdad. On April 20 he "wrote a brief order declaring the coalition to be the 'military authority' in Iraq. [Barbara] Bodine [from ORHA] relayed the order to the State Department and was stunned to learn that it was the first assertion of legal responsibility under the Geneva Conventions. . . . But McKiernan's order was never backed up by Rumsfeld in Washington or Franks in Qatar. It became one of the noble failures in those irretrievable early days."[16]

As then Secretary of State Colin Powell has since summarized CENTCOM's inaction, "We didn't realize we were the government; but we were the government."

Not only was there no real plan for military governance of Iraq, there was an insufficient force structure available to execute any realistic plan that the commanders in theater may have contrived once the need was finally recognized. Troop strength was more than adequate to defeat the Iraqi conventional defense, as Rumsfeld had assessed, but it was completely inadequate for postwar occupation, as General Shinseki had told the president in private meetings and the Senate Armed Services Committee in public hearings the month before the war. By February, however, the train had already left the station. As with ORHA, it was already too late in the game to develop the kind of military governance plan—meaning the mobilizing, organizing, training, and deploying of actual military government and constabulary units—that could have gotten ahead of the events in Iraq in April 2003.

The U.S. Army in 2003 had little institutional knowledge of military governance, did not officially recognize the necessity, and lacked the doctrinal foundation and training base necessary to produce qualified military governance units. It was as if the military governance lessons of the Progressive Era and World War II had been erased from the twenty-first-century American military mind. True, in theater there was the 352nd Civil Affairs Brigade, consisting of two reserve component battalions numbering, altogether, about a thousand soldiers, but it was completely overwhelmed by the enormity of the tasks assigned it.

So the occupying army in Iraq was a thin line of troops well trained and equipped for high-intensity combat but completely unprepared to assume the duties of the fleeing Iraqi government: the ministries, the police, the public works agencies, the courts, the prisons, and the myriad other functions and responsibilities that allow people to live together peaceably. Vast areas and populations were completely beyond American control. Even where Americans happened to be, there were too few translators, too little security, and too little local knowledge to accomplish anything meaningful. Despite everyone's best intentions and efforts, the officers and enlisted men in combat battalions simply lacked the professional backgrounds that would allow them to secure and govern a foreign population on even an interim basis. This inadequate U.S. preparation was experienced firsthand by every Iraqi.

> "We're incompetent, as far as they're concerned," said Noah Feldman, the New York University law professor who went to Baghdad as a constitutional advisor to the Coalition Provisional Authority [CPA]. "The key to it all was the looting. That's when it was clear that there was no order. There's an Arab proverb: Better forty years of dictatorship than one day of anarchy." He added, "That also told them they could fight against us and we were not a serious force."[17]

Indeed, the combat-centric army hardly even recognized that it had a duty to prevent and punish Iraqi-on-Iraqi crimes in the wake of the invasion. As one high-level CPA staffer who served in Iraq in 2003 and 2004 put it:

> We liberated Iraq only to let it rot under the rule of criminals, extortionists, looters, arsonists, murderers, and rapists. . . . It was our lack of power—our inexplicable inability to get things moving, to stop the looters and vandals, to find the troublemakers and punish the terrorists—that led to our being held in contempt by so many Iraqis. The worst part of the aftermath of the war was that we didn't show how powerful we were, but the very opposite; we showed how inept we

were. . . . We proved we had immense destructive power, but little constructive power.[18]

As America failed, anticoalition militias organized and the insurgency began.

Outside of Baghdad, American brigades raced into the provinces with orders to ferret out any remaining Ba'athist resistance and to establish an American presence. These units were, from the Iraqi point of view, the representatives of the new de facto national government, but the combat units that rolled in, indeed the land force at large, often failed to make the right impression during that crucial first encounter. As in Baghdad, U.S. units did not adequately understand their requirements to provide governance under the Geneva Conventions and were unprepared to do so, lacking the planning, force structure, and training necessary to provide governance. The much-publicized Fallujah "incidents" in April 2003 illustrate what can go wrong when post-conflict governance is a strategic afterthought.

Tasked to "assess the current situation and to provide assistance and security as required," on April 23, 2003, the seven-hundred-man 1st Battalion, 325th Infantry Regiment (1/325), from the 2nd Brigade, 82nd Airborne Division, drove into Fallujah, a city in Anbar Province of about three hundred thousand mostly Sunni Arabs, and occupied an empty local school and the deserted Ba'ath Party headquarters building next to the mayor's office, establishing the first American presence in the city. According to battalion commander Lt. Col. Eric Nantz, at this time he had no civil affairs soldiers in support of his unit; the brigade's sole civil affairs team was then located with the brigade headquarters in Habbaniyah.[19] Not only was Lieutenant Colonel Nantz sent to Fallujah without the needed civil affairs expertise, 1/325 also had few Arabic translators, meaning that many of its subunits acted without being able to communicate properly with the Fallujans whom they were to secure.

To be sure, the Fallujans did not want to be occupied. The Iraqi army and Saddam Fedayeen had melted away after the fall of Baghdad and, in the interim, the local tribes and religious leaders had elected a Civil Management Council to replace the Ba'athist government.[20] Almost immediately, the self-liberating (in their view) Fallujans protested and demonstrated against their American occupiers over issues such as the use of the school and the rumors that Americans were using night-vision devices to see through women's clothing. According to Lieutenant Colonel Nantz, there was a "subversive radical element that was able to rally the crowd and then shoot at soldiers."[21]

Five days later on April 28, as 1/325 was preparing to leave Fallujah and turn it over to an element of the 3rd Armored Cavalry Regiment (3rd ACR), a crowd of Iraqi demonstrators approached the American compound. Militants shielded among the Iraqi crowd fired at the American soldiers, who then returned fire. The exact number of casualties is disputed. According to some sources, thirteen

Fallujans were killed and another seventy-five wounded.[22] Lieutenant Colonel Nantz, however, recalls that only five men carrying AK-47s were killed, and although others may have been caught in the crossfire, these high numbers are "not believable." No Americans were wounded or killed in the exchange of fire.

Immediately following the April 28 incident, having spent only six days in Fallujah, 1/325 departed for Baghdad as planned and was replaced by F Troop of the 3rd ACR, which itself was involved in a similar firefight on April 30. Another four Iraqis were killed and fifteen others wounded.

Over the next few months a series of American units rotated through Fallujah; none was ever able to pacify the city. Insurgent attacks continued. Eventually, Americans withdrew from the city, and over the next year and a half the Sunni insurgency in Fallujah would fester and grow into a focal point of the war in Iraq, ultimately leading Americans to depopulate and then clear the city room by room in the massive and destructive assault of November 2004.

Three assessments of American operations in Fallujah highlight the failures of the Phase IV plan. The first observation, made by army officers who served in Anbar, was that the constant rotation of new combat units through Fallujah led to civilian mistrust and was a primary reason that the place "never calmed down" and the insurgency grew.[23]

Equally prescient comments come from the recommendations of the Human Rights Watch report on the April 2003 incidents. The June 2003 report reminded the United States that it is "the occupying power in Iraq" and under international law is "obligated to ensure public order and safety." It further recommended "deployment of adequate numbers of coalition and other military police and constabulary units trained in international law enforcement standards, and provide those units with the resources and equipment to meet Iraq's postwar law enforcement needs." Further, "If there are exchanges of fire, U.S. security forces avoid disproportionate harm to civilians or civilian objects." Finally, "Efforts to enhance communication with local communities should be intensified, starting with adequate provision of translators."[24]

Colonel Nantz's own assessment is that "more training is required, doctrinal change is in order, and combat units need assistance during and immediately following the major combat operations."

None of these assessments or the questions they beg should surprise any student of American military occupations. The inappropriateness of using combat units for governance, hence the need for military governance, constabulary and police units, and the need for ample and productive exchanges between the occupying headquarters and the occupied population formed the foundation of the U.S. Army's World War II–era occupation doctrine and force structure. The army has been rediscovering this knowledge, slowly and painfully, ever since it crossed the Kuwaiti border in March 2003.

Jay Garner's tenure as the occupation leader in Iraq would not last a month. On May 6, two weeks after the ORHA convoy finally made it into Baghdad, President Bush announced that L. Paul Bremer III would become the Coalition Provisional Authority (CPA), the new de facto Iraqi governor.

Ambassador Bremer, a former counterterrorism expert at the State Department who had no apparent expertise on Iraq, was, like Garner, a last-minute choice. It was mid-April, a few days after the fall of Saddam's statue in Baghdad, when administration insiders contacted Bremer about a possible job offer to become Bush's proconsul in Iraq. He spent two weeks in the Pentagon assembling a staff and getting "read in" on the situation in Iraq before arriving in Baghdad on May 12.[25] In his first four days on the job, following plans dictated by Washington, Bremer dismissed all former Ba'athists in the management tiers of the Iraqi ministries, disbanded the Iraqi military, and stopped the formation of a new Iraqi government.[26] As Garner, who was still in Baghdad, later observed, "So on Saturday morning when we woke up, we had somewhere between 150,000 and 300,000 enemies we didn't have on Wednesday morning, and we had no Iraqi face of leadership to explain things to the Iraqi people. We began to pay significantly for those decisions."[27] The dismissals provided immediate recruits, many with military training, to the already growing insurgency.[28]

Bremer's remarkably inept tenure would last until June 2004. The administration strategy had changed from "cut-and-run" to developing Iraq into a democratic, free-market agent of change in the Middle East. Bremer enthusiastically announced, "It's a full-scale economic overhaul. We're going to create the first real free-market economy in the Arab world."[29]

If Bremer's economic plans were naïve and unachievable, his election scheme set the stage for the further division of Iraq along religious and ethnic lines, and an escalating civil war. Against sound advice, Bremer issued CPA Order 96, the new election law, on June 15, 2004, right before he left the country. Rather than hold elections for the national legislature by electing representatives from local districts, thereby making the elected official responsible to a clearly identifiable set of local constituents, as is practice in America and most democracies, Bremer decided that elections would be based on national slates, with each party getting seats in the legislature proportional to its percentage of the national vote.[30]

In January 2005 the first Iraqi election voted in a Shiite majority, which has largely pursued its narrow sectarian interests ever since. This resulted in further disenfranchising the Sunni Arabs at the national level, cementing ethnic and religious divisions, and further fueling the antigovernment insurgency and the sectarian civil war.

Transforming for the Long War

A year after the initial invasion, the army began playing a gigantic game of catch-up.

By early 2004 it became apparent to Gen. Peter Schoomaker, Shinseki's replacement as chief of staff, that the army did not have enough brigades to occupy all of Iraq and accommodate a policy of rotating troops in and out of theater on twelve-month tours. After only a few weeks of discussion, Schoomaker approved a plan to transform the army's thirty-three divisional brigades into forty-three new modular brigade combat teams (BCTs), without an increase in army end strength that Rumsfeld still opposed. However, these new BCTs were not optimized for the stabilization mission in Iraq but were based on a force-on-force paradigm from the late 1990s known as Rapid Decisive Operations.[31]

The U.S. Army's institutional preference is to build units for combat; the noncombat missions that may occasionally be required will be performed by retasked and retrained combat units. For example, when Maj. Gen. Peter Chiarelli and his 1st Cavalry Division arrived in Baghdad in March 2004 for OIF 2, the army had de facto retasked an armored division to perform "stability operations" and work as "glorified policemen and municipal engineers" and "build institutions of local government." By way of training, "Officers . . . had attended seminars conducted by city planners in nearby Austin [close to the division's home base of Fort Hood, Texas]."[32] Realistically, more than a *few seminars* would be necessary to transform soldiers trained to kill enemy into soldiers capable of governing in a strange foreign land. Clearly, America was employing the wrong professional tool for the governance task, but no other had been created. Unlike after World War II, when many U.S. Army combat units were reorganized and retrained as constabulary units, rearmed and reeducated to deal with civilian disturbances and law enforcement, the army has insisted on maintaining a force structure solely devoted to combat and counterinsurgency, the force-on-force fight. To date, the army has not organized a single constabulary unit for occupation duty.

The distinction between constabulary and combat units is important. Young, impressionable soldiers trained to kill the enemy while protecting themselves, the paradigm that necessarily pervades the culture of army combat units, all too often adopt a kill or be killed mentality unsuited to postwar policing and stability operations.[33] The army's preference for "kinetic solutions" in uncertain situations caused many unnecessary civilian casualties, and by 2005 even our coalition allies were complaining in U.S. military journals about the damaging behavior of many U.S. soldiers in Iraq.[34] Because the insurgents and militia members who targeted American soldiers simply blended in with the population, many soldiers began to view all Iraqis as enemies and overreacted to perceived dangers that were not there. (The iconic story has become a soldier shooting up a vehicle that came too close, thinking it was a suicide bomber, only to realize he had killed or wounded an innocent Iraqi family.) Too many Americans became callous to the pain and suffering of the Iraqis whom they were supposed to be assisting, even to the point of being unapologetic for whatever harm they themselves may have caused.[35]

By 2005, especially in the wake of the Abu Ghraib torture scandal, the Haditha massacre, and other reported acts of flagrant misconduct by American soldiers, the army finally began to deliberately train combat units to interact more carefully and respectfully with the Iraqi population. By then it was perhaps a case of too little too late; for a great many Iraqis, the United States had already become the despised occupier. In 2007 an American military police major who had recently returned from Baghdad told me he feared that the U.S. Army had become the new Wehrmacht, professionally excellent military killers but thoroughly mistrusted and hated by the conquered population. He lamented that we were no longer the welcomed citizen-soldier liberators of the World War II era.

Iraqis naturally commingled soldier misbehavior with the "shoot first and ask questions later" attitude of American for-profit security contractors, such as Blackwater, which remained largely unaccountable to either Iraqi or U.S. military authorities. Blackwater's notorious September 2007 killing of seventeen innocent Iraqis in Nisour Square in downtown Baghdad sparked justified global outrage and forced American officials to implement new codes of conduct for contractors in Iraq.[36]

Standing Up the Iraqi Army: Hope and Failure

The American hope in 2005 and 2006 was that the new, rebuilt Iraqi army and police forces would end the need for the large U.S. troop presence. President Bush declared, "Our strategy can be summed up this way: As Iraqis stand up, we will stand down, and when our commanders on the ground tell me that Iraqi forces can defend their freedom, our troops will come home with the honor they have earned."[37]

In the summer of 2004, Lt. Gen. David Petraeus, who had been a division commander in OIF 1, returned to Iraq as commander of the Multi-National Security Transition Command Iraq (MNSTC-I) with the task of training the new Iraqi army and security forces. Over the course of the next year, MNSTC-I appeared to make real progress, greatly expanding the numbers in the various Iraqi Security Forces (ISF) and improving their training and equipment. The Department of Defense optimistically reported to Congress in October 2005 that its program to generate, train, and equip the new Iraqi Security Forces would result in a "completed force generation" of 131,000 Ministry of Defense troops by late 2006 and 195,000 Ministry of Interior policemen by August 2007.[38]

But the force was never as strong as its numbers indicated, and its continued failure to take the lead from American forces demonstrated the lack of trust both the Americans and the Iraqis had in the new security forces. By the fall of 2005, assessments of the Iraqi units were turning from upbeat to negative. Sectarian loyalties, corruption, and apathy pervaded the forces. One reporter assesses, "U.S. trainers have made a heroic effort and have achieved some success with some units. . . . But the Iraqi Security Forces are almost like a black hole. You put a lot in and little comes back out."[39] The mounting violence during the first half of 2006 confirmed the ISF critics' worst fears, and beginning in the summer of 2006 and continuing into the 2007 "surge," American forces had to retake the lead in securing Iraq from the Iraqis.

Companion to the rising unrest in Iraq was a growing army interest in counterinsurgency (COIN) operations, a field virtually ignored in army doctrinal literature since Vietnam.[40] Largely written by Petraeus, Field Manual 3-24, "Counterinsurgency," was published in December 2006, immediately before Petraeus returned to Iraq in 2007 to command the surge. Some in the press publicly extolled FM 3-24 as the breakthrough document that could reverse the course of the war.[41] Clearly, the document was written with Iraq in mind. The opening paragraphs recognize the primacy of governance in COIN warfare. "Political power is the central issue in insurgencies and counterinsurgencies . . . Long term success in COIN depends on the people taking charge of their own affairs and consenting to the government's rule."[42] Later,

> Success in counterinsurgency (COIN) operations requires establishing a legitimate government supported by the people and able to address the fundamental causes that insurgents use to gain support. Achieving these goals requires the host nation to defeat insurgents or render them irrelevant, uphold the rule of law, and provide a basic level of essential services for the populace. Key to all these tasks is developing an effective host-nation (HN) security force. In some cases, U.S. forces might be actively engaged in fighting insurgents while simultaneously helping the host nation build its own security forces.[43]

By all accounts, Petraeus correctly viewed the war as something the Iraqi government and its security forces ultimately need to win; the American military commander can only influence and assist. The key benchmarks became political, not military, and therefore the responsibility of the government of Iraqi prime minister Nouri al-Maliki. Whatever is achieved on the security side can be lost by political influences now largely outside of U.S. control.

Civil War and the Surge

Through most of 2006, Iraq was gripped by a civil war pitting Shiites against Sunnis along the sectarian fault lines in the heartland of Iraq. Baghdad, with its largely intermixed Sunni and Shiite populations, was the scene of nonstop carnage, escalating suicide bombings, and general chaos. Sectarian cleansing campaigns, reminiscent of the ethnic cleansing in the former Yugoslavia, forced millions of Iraqis to flee their homes and abandon their businesses. Insurgent militias and terrorists killed thousands of Iraqis each month, the highest casualty rates since the American invasion.[44] Many more were wounded or injured. With the security forces of the Shiite al-Maliki government openly assisting the Shiite militias, the Sunnis suffered the worst of it, losing entire Baghdad neighborhoods to Shiite gunmen.

The inability, or unwillingness, of the Iraqi government and its security forces to quell the violence appeared to spell defeat for the al-Maliki government and the American effort in Iraq. Ralph Peters, a prominent commentator on military affairs, editorialized in a November 2006 *USA Today* column entitled "Last Gasps in Iraq":

> I supported this war, but the deteriorating situation is starting to convince me that we can't win. Those of us who hoped that the Iraqis could achieve democracy were wrong . . . Iraq is failing. No honest observer can conclude otherwise. Even six months ago, there was hope. Now the chances for a democratic, unified Iraq are dwindling fast. The country's prime minister has thrown in his lot with al-Sadr, our mortal enemy. He has his eye on the future, and he's betting that we won't last. The police are less accountable than they were under Saddam. Our extensive investment in Iraqi law enforcement only produced death squads. Government ministers loot the country to strengthen their own factions. Even Iraq's elections—a worthy experiment—further divided Iraq along confessional and ethnic lines. Iraq still exists on the maps, but in reality it's gone. Only a military coup—which might come in the next few years—could hold the artificial country together.[45]

But the administration did not fold its hand in Iraq, instead opting to surge more U.S. troops into Iraq in a last-ditch attempt to retrieve the situation. General Petraeus was named as the new commander of the military effort in Iraq, charged with a seemingly impossible mission. Immediately upon assuming command in February 2007, Petraeus augmented America's commitment to the Iraqi central government with a three-pronged strategy consisting of aggressive U.S. counterguerilla operations, the formation and legitimizing of nonhostile local Iraqi militias, and local efforts to immediately improve the security of Iraqis in the neighborhoods where they live and empower local leaders. Stopping the sectarian violence was the key, and extremist militias, whether Sunni (al-Qaida or not) or Shiite, were the enemy. If the Baghdad government couldn't govern effectively and ensure the security of the population outside the Green Zone, Petraeus would create and control the tools that could.

The backbone of the surge was the deployment of five additional combat brigades—over thirty thousand additional American soldiers—into Iraq during the spring of 2007. But equally important was the deploying of American units into Iraqi neighborhoods—"where the people slept"—to directly confront insurgent forces and deny them access to and control over the population. Focusing initially on securing Baghdad, U.S. combat forces left their large, fortress-like forward operating bases (FOBs) on the outskirts of the city and occupied, instead, small platoon- and company-size patrol bases in the midst of the capital's embattled neighborhoods. From these bases the Americans would patrol by foot amid the Iraqi population, establishing a degree of contact they never could through vehicle-mounted patrols staging from the FOBs.

The result was an immediate spike in American casualties in the spring of 2007 as the insurgents battled the American forces for population control. American deaths in Iraq rose to 131 in May 2007. But the Americans characteristically won most of the firefights and broke the guerilla's grip on the neighborhoods. Aggressive patrolling began to extend out of the neighborhoods to find and destroy insurgent sanctuaries outside of Baghdad itself. American search-and-destroy offensives reached into the ring of cities surrounding Baghdad where many terrorist operations were based. Focused intelligence and targeting operations against insurgent bases forced the militants to relocate time and again, making them increasingly vulnerable and less effective. Over time, attrition and exhaustion pushed the terrorist threat away from Baghdad. American-led "clear and hold" operations, always augmented by ISF units, pushed al-Qaida back and provided security to local populations, preventing the enemy from returning after the combat units passed through.

Over the course of 2007 and 2008, nearly all of the insurgent strongholds around Baghdad, including the major al-Qaida base in Baquba, fell to the American counterinsurgency offensives. In 2009, al-Qaida in Iraq struggles to maintain its last significant urban base in Mosul.[46] In most aspects, Petraeus' counterguerilla strategy would be familiar to the Americans who fought the Philippine insurgents on Luzon over a century ago.

A second pillar of the Petraeus surge strategy was to legitimize local militias, especially Sunni militias, as long as they protected their neighborhoods and did not engage in terrorist acts against Shiites, ISF, and coalition forces. Often, these militia members were yesterday's terrorists, but given the manifest need of the Sunni communities for protection against Shiite extremist militias, most notably Muqtada al-Sadr's Mahdi Army, and Interior Ministry "death squads," the U.S. decision to allow neighborhood groups to assist in their own security made sense. The American military even paid friendly militia members several hundred dollars a month, helping to raise tens of thousands of militia members willing to work cooperatively with their American defenders and paymasters. The Americans assisted in neighborhood defense by emplacing concrete barriers between rival neighborhoods and allowing militias to man checkpoints, knowing that they would have a better sense of who belonged there and who didn't.[47] Petraeus' motto for this new approach to community policing was "Recruit locally, train locally, deploy locally."[48]

In essence, the United States hedged its bet on the Iraqi central government and the Iraqi Defense and Interior ministries by negotiating and implementing numerous localized security and governmental agreements with various tribes, sheiks, and local civic and business leaders. Further, American unit commanders "partnered" directly with local Iraqi military, police, and militia units for local security, reducing the de facto operational control of the Defense and Interior ministries over their troops in the field. The top-down approach of the early occupation was, in some large measure, replaced by a bottom-up strategy in many insurgent areas.

Of course, all of this became a source of growing discord between al-Maliki, who was popularly viewed as a Shiite sectarian running a largely ineffective Iraqi national government, and General Petraeus, who was a primary architect and the main executor of the new bottom-up strategy.[49] Having once dictated the policies that produced the al-Maliki government, America in 2007 was, in effect, implementing new policies that would at least temporarily reduce its authority throughout Iraq, empowering U.S. and local leaders in their place.

Augmenting Petraeus' enhanced counterguerilla operations was a renewed interest in establishing more effective local governance outside the Green Zone. The United States was finally standing up and funding provincial reconstruction teams (PRTs), effectively bypassing the al-Maliki Baghdad government in an effort to win cooperation in outlying areas.[50] The U.S. Department of State established the PRT concept in 2005, replacing MNSTC-I's ineffective provincial support teams, which had been an underresourced effort hampered by Department of Defense and Department of State turf battles.

Based loosely on the Vietnam War CORDS and the Operation Enduring Freedom (OEF) models, the PRTs were designed to be multinational, interagency, civil-military organizations, roughly one-third military, one-third interagency (that is, nonuniformed U.S. government personnel), and one-third Iraqi nationals or ex-patriots. Unlike CORDS, the formal chain of command for the PRTs runs through the Office of Provincial Affairs to the U.S. ambassador, not through military commanders. Provincial support team lines of effort center around provision of essential services, rule of law, and public diplomacy. By late 2007, ten PRTs were operating throughout Iraq, most co-located with U.S. Army brigade combat teams (BCTs), while additional smaller "embedded" PRTs were being stood up to assist the other BCTs.

Non-Department of Defense (DOD) manning for these organizations was problematic. As of October 2007, none of the PRTs was at full strength, and many had 30 percent or less of the promised nonmilitary personnel. The shortfall was filled by military personnel, often reservists. Overall, the PRT nonmilitary effort was an order of magnitude smaller than the Vietnam CORDS effort, totaling only about two hundred "faces" for about four hundred "spaces."[51] Getting U.S.

government civilians to serve in Iraq outside the Green Zone remained problematic until security improved dramatically.

Attrition Warfare in Iraq

In 2006 and 2007 Operation Iraqi Freedom had become a war of attrition that had reached proportions of demographic significance, especially in the heavily contested areas of the country, Anbar, Baghdad, and Diyala among them. In a free-for-all battle among American-led coalition forces and the three Iraqi factions (Sunnis, Shiites, and Kurds), death and dislocation have ravaged Iraq.

The ease of entrance of the invasion force in March and April 2003, which all told probably cost less than 10,000 Iraqi and coalition lives, was deceptive. As armed insurgents and militias organized against the new imposed governance, the bloodshed and violence grew. Violence begat more violence as sentiments hardened and the aggrieved demanded revenge. High percentages of military-age men joined one armed camp or another and engaged in the fighting, sometimes for noble causes, sometimes not. At the height of the war, hundreds, perhaps thousands, were dying each month, with the exact numbers highly politicized and uncertain. As of July 2007, some authoritative American sources estimated total war-related deaths in Iraq at less 100,000.[52] At the same time, however, Michigan senator Carl Levin put the "slaughter" at "hundreds of thousands of Iraqis."[53] At the higher end of the spectrum, "The Human Cost of the War in Iraq: A Mortality Study, 2002–2006," from the Bloomberg School of Public Health at Johns Hopkins University and the School of Medicine of Al-Mustansiriya University in Baghdad, estimated that 652,000 excess deaths had occurred in Iraq from the time of the U.S. invasion though spring 2006, a figure equivalent to 2.4 percent of the Iraqi population. According to the study, violent deaths numbered 600,000 and were overwhelmingly concentrated among military-age males (fifteen to forty-four years old).[54] American government officials, however, dismissed the report at the time, estimating the total number of deaths in the tens of thousands, but admitted that they did not keep "body counts," a Vietnam-era data point discarded by the army's contemporary leadership. Even higher mortality numbers were reported in an August 2007 poll by Opinion Research Business, a well-regarded British polling company, which estimated that 1.2 million Iraqis had died so far in the war, with a margin of error of 2.4 percent. Nearly half of Baghdad households and one-quarter of households nationwide reported that a family member living under the same roof had died as a direct consequence of the war.[55] This figure, which would represent over 4.5 percent of the Iraqi prewar population, approaches the limits of demographic tolerance.

Also, there is considerable anecdotal evidence of some tribes giving up the fight due to unsustainable casualty rates.[56] Still, with the rapidly growing Iraqi population producing around three hundred thousand new eighteen-year-old males per year, there is little reason to believe that the Iraqi population as a whole is running out of combatants to fuel its largely sectarian civil war.[57]

There is more certainty regarding the war's refugees. An estimated two million Iraqis have fled the country since the American invasion, and in 2007 the number of external refugees continued to increase by forty thousand to fifty thousand per month.[58] Another one to two million Iraqis, depending on which year the counting begins, are internally displaced, often fleeing sectarian violence for the safety of ethnically and religiously homogenous areas. Despite the surge, internal refugees increased by about a hundred thousand per month for most of 2007.[59] The combined refugee figures alone are up to 15 percent of Iraq's prewar population. As of July 2008, few refugees had dared to reclaim their homes.

"Victory" in Iraq

Eight months into the surge, U.S. prospects of "victory" in Iraq remained unclear. The August 2007 National Intelligence Estimate noted in its key findings that:

> There have been measurable but uneven improvements in Iraq's security situation since our last National Intelligence Estimate on Iraq in January 2007 . . . However, the level of overall violence, including attacks on and casualties among civilians, remains high; sectarian groups remain unreconciled; AQI [al-Qaida in Iraq] retains the ability to conduct high-profile attacks; and to date Iraqi political leaders remain unable to govern effectively. There have been modest improvements in economic output, budget execution, and government finances but fundamental structural problems continue to prevent sustained progress in economic growth and living conditions . . . We assess, to the extent that Coalition forces continue to conduct robust counterinsurgency operations and mentor and support the Iraqi Security Forces (ISF), that Iraq's security will continue to improve modestly during the next six to 12 months but that levels of insurgent and sectarian violence will remain high and the Iraqi government will continue to struggle to achieve national-level political reconciliation and improved governance . . . Political and security trajectories will continue to be driven primarily by Shia insecurity about retaining political dominance, widespread Sunni unwillingness to accept a diminished political status, factional rivalries within the sectarian communities resulting in armed conflict, and the actions of extremists such as AQI and elements of the Sadrist Jaysh al-Mahdi (JAM) militia that try to fuel sectarian violence . . . The IC [Intelligence Community] assesses that the emergence of "bottom-up" security initiatives, principally among Sunni Arabs and focused on combating AQI, represents the best prospect for improved security over the next six to 12 months . . .

We also assess that under some conditions the "bottom-up initiatives" could pose risks to the Iraqi government . . . The IC assesses that the Iraqi government will become more precarious over the next six to 12 months . . .[60]

In retrospect, this August 2007 assessment underappreciated the results of the civil war of 2006 and 2007.

It is clear now that the success of the Shiite ethnic cleansing campaigns in and around Baghdad forced the Sunnis to run to the Americans for protection, even if it meant abandoning their erstwhile al-Qaida allies in the process. The so-called "Sunni Awakening" is perhaps a sign that the Sunnis came to realize that they had lost the attrition war to the combined American and Shiite offensive and, like the Germans in World War II, in defeat made alliance with the more generous of their two former enemies. Public opinion polls in Iraq in 2008 showed that the majority of Sunnis wanted continued American presence in Iraq, while the Shiite majority saw no need. With the Sunni threat diminished, Shiite militants began losing power to Shiite moderates. Indeed, religious militants of all stripes seem to be losing popular sympathy throughout Iraqi society. Many news accounts suggest that the religious extremists overreached and are now in disfavor among the populations they relied on for support. The harshness of Islamic law as practiced by al-Qaida in Iraq and the Mahdi Army and a fatigue caused by chronic violence and terrorism, supplemented by the outright criminal conduct of many of the ill-disciplined extremist militias, have driven many Iraqis to seek a more centrist and secular peace. Consequently there is an opportunity to institutionalize a new normalcy should all parties make the efforts to create it.

By the spring of 2008, the surge was unwinding and the five extra American BCTs were beginning to withdraw from Iraq, largely because the army could not continue to deploy troops at the surge pace without breaking the force. In his April 8, 2008, testimony to the Senate, General Petraeus characterized the surge's progress, including the notable reduction in violence, as "fragile and reversible" and could offer the senators no assurance that the American commitment could be reduced below presurge levels anytime soon.[61] American commanders were increasingly convinced that al-Qaida had been defeated in most of its former strongholds, though it remained a threat that could reemerge.

Shiite militias also appear to be on the decline. The al-Maliki government, assisted greatly by U.S. forces, wrested Basra and Sadr City from the grip of Muqtada al-Sadr's Shiite Mahdi Army during its spring 2008 offensives, weakening al-Maliki's militant and radical former political ally.

Though terrorist attacks and other forms of killing occur nearly every day in Iraq, by all appearances the shooting war is abating, and at least for now the risk of renewed civil war similar to what existed in 2006 and 2007 is diminished.

Even in the wake of the January 2009 provincial elections, however, the evidence is slim that the various populations of Iraq are converging on a shared secular identity necessary for a functioning democracy. Six years into the American occupation, Sunni extremists still battle Sunni moderates; Shiite extremists still battle Shiite moderates; Sunnis and Shiites remain largely self-segregated into ethnically cleansed enclaves, often physically protected behind concrete barriers and checkpoints that keep out the hostile militias of the other sect; and the Kurds still aspire to wrestle Kirkuk from the Arabs and expand their autonomous ethnic enclave in the north. It is not now clear that the three major ethnic groups are ready to forgive and forget and move forward with true generosity of spirit. In an area of the world where revenge is a cultural expectation, many fear renewed sectarian bloodshed as the Americans depart over the next two years.[62]

Negotiating a Flawed Peace

In mid-2008, with the worst of the war increasingly seen as over and the end game beginning, the focus in both Baghdad and Washington shifted to the negotiation of a status of forces agreement (SOFA) between the two governments that, beginning in 2009, provides the legal basis for American troops to remain in Iraq.

Clearly, President Bush and his administration were eager to institutionalize an American occupation in Iraq before the Oval Office was turned over to his newly elected Democratic successor, Barack Obama. According, the Bush administration made key last-minute concessions to Iraqi prime minister al-Maliki on troop-withdrawal timelines, legal jurisdiction over American servicemen and contractors serving in Iraq, and the operational freedom of American commanders.[63] The December 2008 agreement requires American troops to vacate Iraqi cities, villages, and towns by June 30, 2009, and "all U.S. forces to withdraw from all Iraqi territory, water and airspace no later than the 31st of December of 2011."[64] Administration spokesmen and some Iraqis remarked that the final departure date could be delayed if both sides agreed. But many Iraqis view the date and the completeness of the American withdrawal as no longer negotiable.

Bush had always argued against withdrawal timelines and accepted them only in the face of Maliki's growing power and intransigence. Assuming that the then ongoing SOFA discussions with the Iraqi government would allow for a lengthy U.S. military presence, in March 2008 the Defense Department awarded $150 billion worth of contracts to a consortium of defense contractors to provide shelter, food, and other services to U.S. forces that would be stationed in Iraq for the next ten years (2008 to 2018).[65] Further, to bind the new Iraqi army to the United States defense-industrial base, the Bush administration steered the Maliki government into contracts for American-built equipment. For example, in January 2009 U.S. officials announced that Iraq will buy a huge fleet of two thousand old Soviet-built tanks, remanufactured and modernized by American defense contractors, under the Department of Defense's Foreign Military Sales program.[66] Bush administration officials and pundits likened the future in Iraq to that of postwar Korea, where tens of thousands of American military personnel still remain over a half century after the conclusion of the Korean War providing the necessary military security behind which democracy could mature and the economy prosper. However, the SOFA agreement, if observed, ensures that Iraq will not be like Korea at all.

The Bush administration's fear that Iraq would become another Philippines, a Republican conquest abandoned by the Democrats once they returned to power, has been at least temporarily assuaged by the Obama administration's February 2009 announcement that American troops levels in Iraq would come down only slightly in 2009, keeping most of the force there to ensure stability for the December 2009 Iraqi national elections, and that there would remain a residual force of thirty-five thousand to fifty thousand training advisors, logistics support, and combat troops after the summer of 2010.[67] Some on the Republican side of the aisle decry this schedule for reduction as being too aggressive, citing the still fragile condition of the Iraqi state and the risks to American interests. On the Democratic side, some urge even more rapid redeployments and deeper cuts in the planned residual force. Still, there seems a consensus that the accomplishments of the war in Iraq can only be secured with continued troop presence.

In light of our SOFA agreement to leave Iraq by the end of 2011, and al-Maliki's insistence that "Iraq will regain independence and sovereignty," our formula for favorably ending the war in Iraq remains as unclear as our reasons for going there in 2003. What were our war aims in Iraq? Iraq was not a threat to the United States in 2001, when the neoconservatives convinced President Bush to invade Iraq as part of the global war on terror, or in 2003, when we finally marched to Baghdad and occupied Iraq. Our trumped-up assertions that Saddam threatened us with weapons of mass destruction or abetted the 9/11 terrorists, if ever credible, were disproved by post-invasion investigations. We may never know the real war aims of that inner cabal of administration officials who manufactured and sold to the American public the case for the invasion of Iraq.

Many observers, in the Arab world, China, and elsewhere, believe we went to Iraq for the oil, which was off-limits to Americans during the Saddam Hussein era. Clearly, the Bush administration had strong links to the oil industry and actively sought to increase Iraqi oil production and provide American oil companies access to Iraqi oil.

Then there is the odd connection between our successful invasion of Iraq in March 2003 and our announcement a month later that we would abandon the Saudi Arabian bases that we had occupied since Desert Shield, a key Osama bin Laden demand of the Saudi monarchy. Indeed, we have relocated those Saudi bases to Iraq and want to keep them there, if not "permanently," then with no firm timetable for removing them. Maintaining strong military forces in the Persian Gulf to secure the vital oil fields and shipping lanes, and increasingly to threaten Iran, is an essential element of American national strategy.

More intriguing still is the connection between a coterie of neoconservative Bush advisors, Israeli desires to solve their Hezbollah problem in Lebanon and Syria, and regime change in Iraq. The "smoking gun" of this conspiracy is the notorious 1996 policy report written for the incoming Israeli prime minister, Benjamin Netanyahu, by a study group headed by Richard Perle that included Douglas Feith and David Wurmser. (Perle, Feith, and Wurmser all became senior Bush administration officials with key roles in formulating Iraqi War policy.) The paper argued that replacing Saddam Hussein's Ba'athist regime with a Jordan-linked Hashemite government would "help Israel wean the south Lebanese Shia away from Hizballah, Iran, and Syria."[68] The desire to strengthen Israeli security was never explicitly cited by President Bush as a reason for America's war in Iraq, yet in the minds of many of the primary architects of Operation Iraqi Freedom, the two ideas undoubtedly commingled and reinforced each other.

All of the unstated and hidden war aims of the various constituencies who supported the war complicate our exit strategy. Iraq has no national interest in solving America's energy problems, basing CENTCOM's air and land forces, or assisting Israel with its security issues. Yet all of these American war aims affect American calculations regarding occupation, withdrawal, and granting the Iraqi government full sovereignty over its affairs. As a result, the proponents of the American occupation seem to desire an Iraq in perpetual limbo—secure enough to allow a substantial drawdown of American forces but never stable enough to allow Iraqis to insist on the complete withdrawal of American forces and their bases, as agreed in the SOFA.

Supporters of the war assure the American people that "troops will be withdrawn as conditions permit," knowing full well that in fractious Iraq no such conditions are possible. Iraq, in essence, becomes a client state, never able to free itself from American protection and, consequently, American demands. Victory in Iraq is not seen as "mission accomplished" and "bring the boys home" but rather as the seizing of a permanent forward operating base from which we can expand our influence and further our interests, publicly stated or not, throughout the Middle East.

Meanwhile, in the United States, Iraq has become the forgotten war. By spring of 2008, American public opinion turned against the war, with about two-thirds of Americans believing that the war in Iraq was a mistake and not worth the effort. About 60 percent favored withdrawing U.S. forces from Iraq at a quicker pace than General Petraeus or the Bush administration then supported, even if the withdrawal adversely affected the stability in Iraq.[69] The financial collapse and recession that engulfed America beginning in September 2008 focused American attention away from Iraq and toward more pressing domestic economic problems. Iraq is now seen as a distraction, not an issue central to America's future.

However, even the "anti-war" party cannot bring itself to abandon the mission in Iraq. Like Vietnam, Iraq has turned into a "tyranny of the weak." Proponents of the American effort there, both in America and Iraq, warn that a steep reduction of American aid and troops will result in renewed chaos and perhaps collapse, ushering in even greater dangers in the region and the world. Even the most "dovish" in Washington are fearful of being labeled as the defeatists who didn't "support the troops," who "lost Iraq," and are therefore responsible for all the future evils in the Middle East. Consequently, Iraq has come to possess America, and it is unclear how we will ever free ourselves from its grip.

No war in American history has been as confused in its purpose and execution as the war in Iraq. The effort has been unworthy of a nation that proclaims itself the sole world superpower and supports a costly professional military unrivalled in the world. Iraq refutes the paradigms of warfare that became vogue in the U.S. military in the wake of the Cold War. Our inability to conceptualize the political and military requirements of our offensive war in Iraq, a product of flawed doctrine and wishful thinking, led to tremendous failures of prediction and planning that continue to undermine our possibilities in Iraq and, indeed, throughout the Middle East. The American people and the Iraqis themselves deserved a better effort, but during the critical two years from late 2001 through 2003, the army as an institution failed to understand its business and, consequently, steered the nation and itself into a quagmire in Iraq. Our country has been paying for the army's failures ever since.

It is now uncertain whether future good efforts by more enlightened military commanders and political leaders can redeem our investment in Iraq. So much of the fate of our enterprise in Iraq will be determined by the Iraqis, not by us. What is more certain is that the military must critically assess the deficiencies of its Full-Spectrum Operations doctrine and rediscover ways of thinking about war that are more likely to lead to the correct and timely decisions necessary to achieve victory and avoid folly. A retreat to pre–Cold War thinking would be a step forward.

Chapter 8

America's War in Afghanistan

Afghanistan is one of the poorest and least-developed nations on Earth, rivaling Haiti and the worst areas of sub-Saharan Africa. Afghanis have never produced an economic surplus sufficient to support the various hierarchies and bureaucracies required by centralized government. A subsistence farming society that produces no significant taxable income cannot afford nationhood as it is currently understood and practiced in the West.

An inadequate tax base results in no meaningful governmental services, which undermines the very legitimacy of government and the very notion of nationhood. In 2006 the domestic revenue base of the Afghan government was only thirteen dollars per capita, less than four cents a day, wholly insufficient to provide roads, schools, police, a judiciary, water, electricity, or any of the other services that we in the West associate with governance.[1]

In Afghanistan, as in other destitute societies, tribalism has prevailed as a logical response to extreme human poverty, tribal governance being the simplest and lowest-cost solution to resolving the immediate disputes and problems that arise among people living a hand-to-mouth existence in small, geographically isolated communities.

Unsurprisingly, Afghanistan's history is one of strong tribes and subtribes (clans) and weak central authority. NATO's former International Security Assistance Force (ISAF) commander in Afghanistan, Gen. Dan McNeill, recalled in a recent address that Afghani tribalism had forced Alexander the Great into a seven-year-long counterinsurgency war nearly 2,300 years ago. The irreconcilability and incompatibility of the European nation-state and the Afghani tribe, evident even then, continues to this day.[2] Whatever central governments emerged from time to time in Afghanistan were essentially weak confederations of tribes, each tribal leader retaining the strength and prerogative to withdraw from the loose association if it no longer suited his interest.[3] Tribalism had sufficient strength to defeat central governments and invaders alike and left Afghanistan a backward nation in the remotest recesses of central Asia.

The War Against the Soviet Invasion

In 1978, the Soviet Union sought to communize Afghanistan and install a strong government in Kabul. The Communist coup d'état inspired immediate armed resistance from the Afghani tribes and the Muslim fundamentalist community, which became inspired by the success of the Iranian revolution next door. By the end of 1979, with the Communist government in Kabul clearly failing, the Soviet Union invaded Afghanistan, emplaced its own indigenous Communist leader, Babrak Karmal, and took control of the counterinsurgency efforts. Immediately, many in the American government sensed an opportunity to bleed the Soviets by supporting the Afghan insurgency, and within a matter of days President Jimmy Carter authorized the CIA to supply the mujahedin guerilla forces with weapons and other support.

For many in the CIA, the sole goal was to kill Russians to avenge Soviet support of the Communist Vietnamese a decade earlier. The CIA did not initially consider that the mujahedin might actually defeat the Soviets, so no attention was paid to possible governance consequences of the new war in Afghanistan. For ease of delivery and administration, the CIA agreed that the American aid would be disbursed to the Afghanis directly by the Pakistani army's Inter-Services Intelligence (ISI), a secretive, antidemocratic agency with its own Islamist agenda.[4] Further, the Americans secured matching financial assistance from the Saudi Arabian government, which in turn secured additional private donations from other Saudi sources, many with strong Wahhabist leanings.

Suddenly, hundreds of millions of dollars were flowing into the Pakistani ISI, and it used the dollars to pursue not just the anti-Soviet war in Afghanistan but also the struggle against India in Kashmir and the expansion of an Islamic fundamentalist base in the tribal areas along the Durand line, the recognized, albeit popularly ignored border between Afghanistan and Pakistan.

Not only did the ISI funnel most of the money and arms to Islamist warlords and tribal leaders, including Gulbuddin Hekmatyar, it also tended to freeze out more traditionalist tribal leaders, such as Ahmed Massoud, and royalist Pashtuns.[5] Large amounts of Saudi oil money poured into the tribal regions to build thousands of madrasses to expand fundamentalist religious education and grow the mujahedin recruiting base. It is likely that Osama bin Laden acted as a liaison between Saudi intelligence agencies and the growing Islamist educational and military establishment operating out of northwestern Pakistan.[6]

The Red Army's heavy-handed, scorched-earth campaign succeeded in inflicting large numbers of casualties and widespread destruction, but during its first four years it made only limited progress in securing the countryside. By the beginning of 1984, the Soviets, with over one hundred thousand troops in Afghanistan, had achieved little more than a strategic stalemate against the two hundred thousand or so mujahedin fighters with their CIA-furnished small arms.[7] The CIA estimated that their Afghani surrogates had killed or wounded about seventeen thousand

Soviet soldiers and controlled 62 percent of the countryside. The Red Army had lost hundreds of aircraft and ten thousand or more armored and transport vehicles. Soviet financial costs had reached about $12 billion, a tremendous return on the CIA's $200 million in assistance to the mujahedin to date.[8]

Success bred greater ambition, and over the course of the next year the American government decided to double and then redouble its aid to the mujahedin in hopes of actually driving the Soviets out of Afghanistan and inflicting a Vietnam-like defeat against our Cold War foe. Accordingly, Congress appropriated $250 million of support for fiscal year 1985, later $470 million for fiscal year 1986, and $630 million for fiscal year 1987, all of which was matched by the Saudi government and further augmented by Saudi "unofficial" donations.[9]

Regrettably, although characteristically, the American policymakers, who were fixated on the force-on-force battle, paid virtually no attention to the governance consequences of the change in policy. Now seeking regime change through offensive war, America's policymakers failed to think through what kind of new government our efforts should promote. Dismissing Afghanistan as essentially ungovernable, the CIA was content to funnel the massive American aid through the ISI, which directed it to the most Islamist of the various mujahedin factions, who were by now building a power base in the border areas, alarming both the Soviets and the more moderate and traditionalist Afghani population. Thousands of foreign Arab jihadists began to accompany the weapons flow.[10]

In March 1985, Mikhail Gorbachev, a dovish economic reformer preaching "openness" and "reconstruction," assumed the reins of power in war-weary Russia. Initially unwilling to admit defeat in Afghanistan, he instead decided to give his generals one to two years to end the war favorably and virtual carte blanche permission to escalate the war as needed.[11] Babrak Karmal was sacked as ineffective, replaced as head of the Soviet-supported government by the ruthless Afghan secret police chief Mohammed Najibullah, who, with KGB assistance, had greatly improved the Afghani intelligence services. Additionally, the Soviets deployed larger numbers of elite Spetsnaz forces and Mi-24D Hind attack helicopters with improved armor to defeat mujahedin small-arms fire.[12]

The renewed and escalated Soviet offensive, which coupled new weaponry, units, and tactics with widespread scorched-earth and terror policies, inflicted severe casualties on the Afghan resistance and the general population. By the summer of 1986, CIA and Pakistani analysts actually feared that the new Soviet assault might succeed in tipping the balance against the mujahedin. To save the day, the CIA began shipping the first of about two thousand man-portable Stinger antiaircraft missiles to the mujahedin.[13] The state-of-the-art infrared homing missiles were more than a match for the Soviet Hind attack helicopters and fixed-wing aircraft, and, once delivered in sufficient numbers, checkmated the new Soviet heliborne tactics. Indeed, the Stingers proved so effective that American policy in Afghanistan would soon center around buying back unused missiles from our former "allies," who, after the Soviet withdrawal in 1989, began to look more like "terrorists" who might use them against commercial aircraft.

In April 1988 Gorbachev agreed to withdraw all Soviet troops from Afghanistan by January 1989, though retaining the right to assist the Najibullah regime with military and other aid. The Soviets withdrew as promised, ending their nine-year-long counterinsurgency effort.

Soviet scorched-earth policies had caused staggering levels of death and destruction. By 1990 there were 1.3 million Afghani dead (8.4 percent of the 15.5 million prewar population) and 3.5 million refugees (23 percent), and half of Afghanistan's thirty thousand villages had been destroyed. The persistence of the Afghanis in their protracted struggle for home rule in the face of such severe losses is indeed remarkable. In contrast, Red Army casualties were fewer than fifteen thousand dead and about fifty-one thousand wounded, the equivalent of a bad day's losses in World War II.[14] Ultimately, the Russian people rejected their offensive war against Afghanistan, despite its relatively minor cost, in a way they never could have dismissed their defensive war against Nazi Germany.

The Taliban Era

Americans and Pakistanis alike believed that Kabul and the Communist regime would fall as soon as the last units of the Red Army withdrew, but such was not the case.[15] Torn between the devil they knew, Najibullah, and the devil they feared, the ISI-supported Islamist armies of Hekmatyar and others, the various tribes and warlords recombined again and again to produce a sporadic stalemated civil war that ended, using the term loosely, only with the Taliban's capture of Kabul in September 1996 and the imposition of its radical Islamist agenda, including support for Osama bin Laden's newly formed al-Qaida. Only the Northern Alliance, a mostly Tajik moderate group north of the Hindu Kush, resisted Taliban rule into the new millennium.

American neglect clearly tipped the balance of power in Afghanistan against our long-term interests. With the Soviet withdrawal, Afghanistan's governance became a third-tier issue for American policymakers. CIA funding for the mujahedin dropped to $280 million for fiscal year 1990 and stopped altogether on January 1, 1992. Meanwhile, Saudi financial support for Islamist forces in Afghanistan remained high.[16] With the Red Army defeated and withdrawn, America quit funneling money to Afghani centrists, abandoned its Kabul embassy, and neglected to assign ambassadors or even CIA station chiefs to Afghanistan.[17] Edmund McWilliams, the Department of State special envoy to the Afghan Resistance, wrote regarding America's unconcern in 1992:

> [The] principled U.S. posture of letting Afghans find solutions to "their problems" fails to take into account a central reality: Intense and continuing

foreign involvement in Afghan affairs—by friendly and unfriendly governments and a myriad of well financed fundamentalist organizations—thus far precluding the Afghans from finding "their own solutions." [The hands-off policy of the United States] serves neither Afghan interests or our own . . . The absence of an effective Kabul government has allowed Afghanistan to become a spawning ground for insurgency against legally constituted governments. Afghan-trained Islamic fundamentalist guerillas directly threaten Tajikistan and are being dispatched to stir trouble in Middle Eastern, southwest Asian, and African states.[18]

Operation Enduring Freedom (OEF)

American neglect of affairs in Afghanistan ended on September 11, 2001, when Afghan-based al-Qaida terrorists used four hijacked planes to destroy the World Trade Center in New York and damage the Pentagon in Washington, killing over two thousand Americans.

America's war in Afghanistan, Operation Enduring Freedom (OEF), began on October 7, 2001, when U.S. special forces and CIA teams, numbering about a thousand, teamed up with Northern Alliance and other anti-Taliban militias and began directing U.S air power against Taliban targets. The Taliban regime crumbled spectacularly, losing Mazur-e-Sharif in the north on November 9, fleeing Kabul on November 12, and abandoning its southern and seminal Pashtun base at Kandahar on December 9, 2001.[19]

Early American reluctance to commit infantry units in the Afghan campaign, an outgrowth of Defense Secretary Rumsfeld's desire to demonstrate his defense "transformation" away from conventional ground forces, prevented the U.S. Army from surrounding and annihilating much of the Taliban leadership, al-Qaida, and bin Laden himself at Tora Bora in early December. Most of our worst enemies exfiltrated through the snow-covered high-mountain passes to their old refuges inside Pakistan.[20] Allowing the al-Qaida nucleus to escape capture or death was a tragic strategic error that would, as a consequence, cause our lightning campaign to bring al-Qaida to justice to escalate into a decade-long, multi-trillion-dollar global war on terror and invasion of Iraq.

Still, in the operational victory that was the opening two months of the war in Afghanistan, the small American effort had toppled an enemy regime and driven a terrorist group away from its training camps and institutional base. Only a handful of American servicemen died in the effort.

Establishing a New National Government: Mission Impossible

The 2000 Bush election campaign had promised not to mire the country in "nation-building" projects overseas. It's not surprising, then, that the administration desired to leave as small a military footprint as possible in Afghanistan. The Bush team accordingly sought to internationalize Afghanistan's political reconstruction.

On November 14, 2001, immediately after the fall of Kabul, U.S. diplomats pushed Resolution 1378 through the UN Security Council. The resolution called for the United Nations to have a central role in "establishing a transitional administration and inviting member states to send peacekeeping forces to promote stability and aid delivery." Consequently, the UN invited various Afghan parties to a conference in Bonn. The conference resulted in an agreement to form a thirty-member interim administration to rule until June 2002, when a *loya jirga*, a traditional meeting of tribal and factional leaders, could meet to chose a permanent government and establish the mechanism for creating a constitution and holding national elections. The agreement was endorsed by UN Security Council Resolution 1385 on December 6, 2001. Hamid Karzai, a Pashtun diplomat and U.S. supporter, was made the interim president. This choice was ratified by the *loya jirga* when it met later. Subsequent milestones include the constitutional *loya jirga* that ratified a new constitution in January 2004, the presidential election in October 2004 that elected Karzai, and the September 2005 parliamentary elections.[21]

Efforts of the United States and United Nations to establish a strong central government in Kabul have not achieved the success that the chronology above might imply. The reach of Karzai's government does not extend far outside the capital city of Kabul, where it enjoys the protection of ISAF, a NATO force of about thirty-one thousand soldiers pledged by some thirty-five countries. Efforts to expand central authority into the countryside have yielded meager results. A November 2006 CIA assessment found that "increasing numbers of Afghans view the Karzai government as weak and corrupt.[22] He is often mocked as the "Mayor of Kabul." Lacking real resources, however, Karzai is incapable of providing real governmental services. Unable to provide real services, especially security and economic development, it is unsurprising that large numbers of Afghanis question the government's legitimacy and good intentions.[23]

Recognizing the need to extend governance and governmental services outside of Kabul, the U.S. military began establishing PRTs in late 2002 and early 2003, echoing the CORDS thinking of the Vietnam era. By June 2005 the United States had created thirteen PRTs, with an additional nine operated by our NATO allies in ISAF. The U.S. PRT model has eighty-two personnel, almost all military, with a commander and staff, civil affairs teams, a military police unit, a psychological operations team, an intelligence team, medics, an infantry platoon, and support personnel. Each team is provided a State Department representative, a USAID representative, a U.S. Department of Agriculture representative, and a representative from the Afghanistan government's Ministry of the Interior.[24]

In reality, unfortunately, the representatives of the U.S. civilian agencies often arrived late, if at all, and lacked sufficient experience, language skills, and cultural knowledge to be truly effective.[25] The vulnerability of the PRTs to insurgent attack forced them to collocate with larger military units on their forward operating bases. The military "combatant" character of the PRTs scared away many of the aid-giving nongovernment organizations (NGO)—for example, Doctors Without Borders—which profess a need for impartiality to gain access to those in need. The security requirements inherent to stability operations are largely military, and the perception that the U.S military, not to mention the U.S. government, is a combatant and an occupier pushes NGOs to maintain independence from the PRT effort.[26]

Provincial reconstruction team commanders were charged with promoting Afghan central government authority, but they often found themselves, for practical reasons, supporting the agendas of local and provincial leaders.

In many cases, however, PRT support for local leaders was counter-productive. A number of provincial governors and police officials were old-line warlords, militia commanders, or regional power brokers whose loyalties were questionable and whose interests diverged from those of the central government. Support from the PRTs actually enabled these leaders to further distance themselves from relying on the central government.[27]

Indeed, the very magnitude of U.S. and international aid flowing into Afghanistan serves to underscore the frailty of the Afghan government. Too often the bottom line is that the Americans have money and the Afghan government lacks it, hardly a formula for building governmental legitimacy. Internally raised government revenues of about $450 million per year are an order of magnitude less than American and international contributions.[28] Worse still, at least half of Afghanistan's gross domestic product derives from illegal heroin production and trafficking, and all of these proceeds go to antigovernment factions, including the Taliban.[29]

The Taliban Regroups in Pakistan

Left alone to sort out its internal problems with the beneficent assistance of the West, Afghanis could perhaps envision a gradual, heavily subsidized evolution toward a coherent state. However, such a dream would ignore the very real threat from a resurgent Taliban operating with near impunity from their safe havens in Pakistan, supported still by Islamic extremists from Saudi Arabia and elements within the ISI. The Pashtun populations on the Pakistani side of the Durand line occupy a no-man's-land ungoverned by either Kabul or Islamabad, which, willingly or not, has become the headquarters of al-Qaida and other extreme Islamist entities.

In September 2006 the government of Pakistan signed a peace deal with the "tribal elders of North Waziristan and local mujahideen, Taliban and ulama [Islamic clergy], codifying the de facto safe-haven enjoyed by the enemies of the Karzai regime."[30] As of late 2006, Mullah Umar, the leader of the Taliban and the former ruler of Afghanistan until the American invasion of 2001, operated from the Pakistani city of Quetta, immediately across the Durand line from his former base of operations at Kandahar.[31] Meanwhile, Karzai's Islamist rival and former ISI favorite Hekmatyar is rumored to be operating out of his twenty-year-old base of operations in Peshawar, on the east side of the Khyber Pass. Bin Laden may be there also.[32] In between, in the border area called South Waziristan, a new Taliban leader, Baitullah Mehsud, who has been blamed by Pakistani and U.S. authorities for a succession of attacks, including the 2008 assassination of returned former Pakistani prime minister Benazir Bhutto, battles the Pakistani army for control of that province.[33] American armed Predator drones and other aircraft now regularly conduct precision strikes against suspected al-Qaida and Taliban targets on the Pakistani side of the Durand line.

It is supreme irony that the Islamist base in the tribal border areas that the United States substantially created in the 1980s to defeat the Soviets in Afghanistan now threatens not only the government we have subsequently installed in Afghanistan but also the stability of Pakistan itself and its recently developed nuclear arsenal. Driven out of Afghanistan by American counterinsurgency efforts and harassed in the border regions by covert American strikes, the Taliban, al-Qaida, and their kindred Islamic extremists expand their range of operations ever southeastward from their mountain isolation toward Islamabad and other Pakistani lowland population centers. In 2008 the Taliban defeated the Pakistani security forces in the Swat Valley, only eighty miles from the Pakistani capital, and imposed Sharia (Islamic religious) law in the region of over a million people. In February 2009 the Pakistani government announced a truce with the Taliban in Swat, in effect admitting that it was too weak to retake control of the valley. Meanwhile, cadres of Pakistani Taliban, many trained in the border area madrasses that the Saudis began funding with American consent in the 1980s, are now finding their way into Karachi and the Punjab, with aims to conquer all of Pakistan.[34] Many American officials are now contemplating the unthinkable: the collapse of the fragile Pakistani state and the chance that nuclear weapons and delivery systems could fall into Islamist hands. Nothing we could gain in Afghanistan is worth what we could lose if Pakistan disintegrates.

Escalating the War in Afghanistan

Indeed the security challenges in Afghanistan and neighboring Pakistan are forcing the U.S. Department of Defense to pour ever more troops into Afghanistan. From its small U.S. Army special forces beginnings, American troop commitment had grown to over 25,000 men by the end of 2007 and 36,000 troops by the January 2009 change of administrations. (The U.S. goal had been to reduce force levels to less than 17,000 by the end of 2006, but the goal was scrapped due to the Taliban resurgence.[35]) Over 20,000 non-U.S. troops under ISAF command augment the U.S. force. The Afghan National Army, in the process of being rebuilt, rearmed, and trained by U.S. and ISAF forces, numbered about 36,000 at the beginning of 2007 and has since grown to around 80,000, with plans to increase to 135,000. The Afghan National Police number about 70,000, of which 50,000 are trained and equipped.[36] After his mid-2006 visit to Afghanistan, Gen. (Ret.) Barry McCaffrey, a key U.S. advisor on military matters, characterized the state of the Afghan security forces as follows:

> The Afghan Army is miserably under resourced . . . This is now a morale factor for their soldiers. They have shoddy small arms—described by Defense Minister [Abdul Rahim] Wardak as much worse than he had as a Mujahideen fighting the Soviets 20 years ago . . . Afghan field commanders told me they try to seize weapons from the Taliban who they believe are much better armed . . . [The Afghan National Police] are in a disastrous condition, badly equipped, corrupt, incompetent, poorly led and trained, riddled by drug use and lacking any semblance of . . . infrastructure.[37]

Indeed, concerned about the poor security offered by the Afghan National Army and Afghan National Police, British military commanders in Afghanistan are pushing for the creation of tribal militias to provide local security where the national army and police are incapable. British prime minister Gordon Brown stated that his forces would provide "support for community defense initiatives, where local volunteers are recruited to defend homes and families modeled on traditional Afghan arkbaki [village militias]."[38] American officials debated and rejected a similar idea in 2004, believing that the creation of centrally controlled national forces, built along Western models, provides the best path for Afghanistan. Today, the Americans are launching pilot programs aimed at producing "local protection forces" along the lines the British suggested.

Despite the plethora of security forces, America, now in the eighth year of the war, believes it is losing in Afghanistan. Attempting to "stabilize a deteriorating situation in Afghanistan," the new Obama administration ordered seventeen thousand more troops to the country, a deployment that will bring the U.S. force total to well over fifty thousand by the summer of 2009.[39] Meanwhile, the popularity of the American effort and the Karzai government among the

Afghan population plummets, with a quarter of the Afghan population now believing that attacks on U.S. forces are justifiable.[40] A U.S. Army general officer in Kabul recently stated that the effort to build Afghani national governance suffers from a lack of human capital and a culture of corruption. "Linkage" between the national government in Kabul and the local governments in the outlying areas is weak to nonexistent. The Afghan National Police, who have traditionally exploited and abused the population, still lack the "protect and serve" ethos that Westerners would expect from a police force. Afghanis view the national police unfavorably, preferring traditional, tribal, often Taliban solutions to the American initiatives to nationalize law enforcement and justice.

Reassessing Our War Aims

Some American pundits despair that the war is already lost. As commentator Ralph Peters wrote in the wake of Obama's announcement of the troop surge:

> The conflict in Afghanistan is the wrong war in the wrong place at the wrong time. Instead of concentrating on the critical mission of keeping Islamist terrorists on the defensive, we've mired ourselves by attempting to modernize a society that doesn't want to be—and cannot be—transformed . . . Expending blood and treasure blindly in Afghanistan, we do our best to shut our eyes to the worsening crisis next door in Pakistan, a radicalizing Muslim [*sic*] state with more than five times the population and a nuclear arsenal. We've turned the hose on the doghouse while letting the mansion burn . . . Initially, Afghanistan wasn't a war of choice. We had to dislodge and decimate al-Qaeda, while punishing the Taliban and strengthening friendlier forces in the country. Our great mistake was to stay on in an attempt to build a modernized rule-of-law state in a feudal realm with no common identity. We needed to smash our enemies and leave. Had it proved necessary, we could have returned later for another punitive mission. Instead, we fell into the great American fallacy of believing ourselves responsible for helping those who've harmed us.[41]

Indeed, little positive has been accomplished in the eight years since the American invasion that would survive our departure. Whether the American forces can succeed in building a strong central authority in Kabul that can subordinate the traditional power and prerogative of the various tribes and ethnic groups, despite all the historical precedents of failure, is the proposition now being tested. As former ISAF commander McNeill so accurately phrased it:

> [N]one of the necessary conditions for redefinition of a civilization are evident in Afghanistan: As such, its transformation is likely to be

prolonged and painful . . . Our strongest hope is that Afghans will develop a strong sense of nationhood and a new will to provide for the common good. This demands that Afghans develop new concepts for association among the tribes and a new identity that transcends tribalism . . . [A]ny notion that we can "fast track" a nation-in-crisis toward a more modern form of government is wrong.[42]

Our lightning military conquest of 2001, almost over before it began, was only a minor first achievement in what the military and the nation have now come to realize is surely among the most ambitious, perhaps foolhardy, attempts at governance building ever attempted.

Chapter 9

The Military's Vacuous New Doctrine

The army writes doctrine primarily to explain how wars are won. The doctrine has two intended audiences. The most obvious is the army itself; doctrine gives us a common understanding of what we're doing. The second, no lesser audience is the civilian leadership to whom we are responsible; our doctrine expresses to them the requirements and realities of the wars that they may direct us to fight. Hard realities govern our estate; to think otherwise puts our soldiers, allies, and nation at risk.

Both audiences deserve doctrine based on the best available evidence. Our doctrine should include only those statements and ideas that are supported by the fairest reading of the best available data, both historical experience and, judiciously, the results of forward-looking experiments and exercises. Doctrine should not be a repository of intellectual fad, unproven assertions, or wishful thinking. Nor should doctrine avoid key unpleasant truths in an attempt to make war seem bloodless or agreeable. At our current historical juncture as the world's preeminent military power, we need to accurately state to the civilian leadership how military action contributes to a prosperous and peaceful world for ourselves and all humanity.

Were our strategic military policy purely defensive, meaning that we only wished to preserve our own way of life for ourselves and our allies, our doctrine would not be in crisis. Our overwhelming military strength, in addition to our political and economic power, would easily overwhelm any nation aggressing against us. Rather, it is our offensive military doctrine that has failed us. We simply cannot, as we once could, manage to impose new governance on nations far weaker than we. This is a tremendous failure at a time when many of our national leaders are calling for offensive wars to democratize backward states, or to replace unfavorable regimes, or to permanently preempt and prevent terrorist strikes against us. In the 1990s and early 2000s when the neoconservatives and others began calling for a new American strategic offensive, one would have expected that the army would "dust off" its offensive war doctrine to ensure that it could meet the more aggressive demands of the emerging new leadership. One would have assumed that the military would have looked at its own history, the recent world experience in offensive war, the state of technology, and the state of global military forces and update accordingly its thinking on the waging of successful offensive war. But we didn't, and consequently the nation pays a bitter price in Iraq and Afghanistan.

How did we lose our way?

In all fairness to the army's current generation of doctrine writers, the army has never in its history written down and codified a coherent vision of warfare. Prior to World War II, the small peacetime army never established a "doctrine" bureaucracy the equivalent of today's Training and Doctrine Command. The wartime army was always too busy with the practical tasks at hand to record in detail the reasons it did what it did. The American military favored fighters, not writers, and this led to a paucity of the kinds of written documents—professional journals, doctrinal publications, and training manuals—upon which we've recently grown dependent. As Russell Weigley remarked in 1973 in his authoritative *The American Way of War*, "[T]he evolution of American strategy before the 1950s has to be traced less in writings about strategy than in the application of strategic thought in war. It has to be a history of ideas expressed in action."[1]

Lack of published theory seemed not to harm the army's war efforts. The institutional belief in the primacy of attrition warfare allowed for an immediate concentration of intellectual effort on how the attrition would be achieved: the tactics, the operations, the units, and the equipment that in their accumulated weight would crush an opposing army, indeed an opposing nation, over the course of a war. Establishing an overwhelming strategic base for the attrition fight, the critical element in victory, was always the crucial first step. The details of the ensuing campaigns, important as they may eventually become, could be decided while the country mobilized.

In a military that currently recognizes different levels of military thought—strategic, operational, and tactical—the army's published doctrine for the past century has started with operations and devolved into more detailed discussions of tactics, with strategy being skipped altogether. Since World War I the army's capstone doctrinal manual has always been titled Operations, an intellectually odd point of entry. This sort of intellectual sloppiness went unnoticed until the 1970s, perhaps because a strategy of attrition warfare was always assumed. Our defeat in Vietnam caused the U.S. military to abandon our historical preference for attrition warfare. Body counts, supremely important indicators of attrition, fell into disfavor in the denouement of the Vietnam War. So, too, the draft was discarded in favor of a smaller professional force, prompting a rethinking of the attrition strategy.

As discussed previously, the army solved its newfound intellectual deficit, at least in part, by embracing Clausewitz's *On War* as its "new" unofficial theory of war and strategy. Still, the post-Vietnam army never wrote its own doctrinal manual on war or strategy writ large.

Neither has the joint staff, which, since the 1986 Goldwater-Nichols Act, is the agency responsible to think at the strategic and operational levels. When that 1986 legislation relieved the army of its war-fighting mission, that function now being invested in the joint regional combatant commanders, the army simply no longer had the authority to write strategic or operational war doctrine for the joint force and the nation. Unfortunately, the joint doctrinal publications, like the army field manuals, begin their discussion of war with operations, disembodied from a stated theory of war or strategy. Of course, the "joint staff" is not an armed service of its own. It is made up of soldiers, sailors, marines, and airmen. As the primary land warfare service, the U.S. Army is well represented on the joint staff and on the staffs of the joint regional commands. But although the army can influence joint doctrine, it does not write it. The joint doctrine now being produced is essentially a political compromise between the various interests of the various services. Consequently, nothing in joint doctrine is controversial or incisive.

Attrition Warfare and Military Governance

With regard to offensive war, American military doctrine reached its zenith in World War II and has been in decline ever since. World War II was the last time the army had a clear understanding of both attrition warfare and military governance, which the evidence indicates are the two crucial elements for success in offensive war. It is also the last offensive war in which the government provided the military with the necessary resources to perform both missions.

This coincidence is no accident. Had the army's World War II doctrine stated its inability to win a war of attrition or to successfully occupy defeated enemies, the nation would have fought the war differently. But the World War II army warfighting doctrine, Field Manual 100-5, "Operations," is consistently confident that army operations would end in "the destruction of the enemy's armed forces in battle" and that the "purpose of offensive action is the destruction of the hostile armed forces."[2] There is no mention of center of gravity, but rather an underlying belief that the army's superior firepower and mobility would destroy the enemy and eventually force him to surrender. As Patton so eloquently summarized our World War II doctrine, "Battles are won by frightening the enemy. Fear is induced by inflicting death and wounds on him. Death and wounds are produced by fire. Fire from the rear is more deadly and three times more effective than fire from the front, but to get fire behind the enemy, you must hold him by frontal fire and move rapidly around his flank . . . Catch the enemy by the nose with fire and kick him in the pants with fire emplaced through movement."[3] Maneuver is seen as valuable because it affords an opportunity to inflict casualties upon the enemy at a higher rate than otherwise possible and avoid the friendly casualties inherent in the more direct approach.

Of course, the army also invested heavily in mortars, artillery, and airplanes to ensure unrelenting fire on the enemy from above throughout the depth of the battle area. Even in his homeland the enemy would find no sanctuary from America's wrath. Our doctrine and practice ensured that any enemy anywhere would fear the effects of American firepower. Battle casualty statistics from both the European and Pacific theaters validated the American doctrinal formula.[4]

But the War Department knew that the destruction of the German and Japanese military through offensive action was not a sufficient basis for building a permanent peace and therefore promulgated Field Manual 27-5, "Military Government," which laid out the policies, organizations, and responsibilities for replacing the hostile regimes of our defeated enemies with friendly and democratic governments of our own making. Collectively, these two documents provided the intellectual capital behind the army's successful prosecution of World War II, animating, as they did, U.S. Army force structure and training.

The first doctrinal pillar supporting the American way of offensive warfare to fall was military governance. Field Manual 27-5, first published in 1940, revised and retitled as the "United States Army and Navy Manual of Military Government and Civil Affairs" in December 1943, was last published in a 1947 edition that is now obsolete, as is the successor 1957 Army Field Manual 41-10, "Civil Affairs Military Government Operations." Since the late 1950s, the army has lacked the doctrine and force structure necessary to successfully occupy and reconstruct defeated nations.

Army force-on-force doctrine changed little from the 1940s through the AirLand Battle doctrinal manuals of the 1980s and 1990s. For instance, the 1968 Field Manual 100-5 discusses the spectrum of Cold War conflict, from nuclear war to limited war, but the stated principles of war are those adopted by the army after World War I. The 1973 version, which emphasized defense rather than offense, was more of a weapons employment manual addressed to the lowest tactical levels; it extols the salutary effects of firepower in the American way of war. "The skillful commander substitutes firepower for maneuver whenever he can do so . . . Massive and violent firepower is a chief ingredient of combat power."[5]

The major innovation of the 1980s AirLand Battle doctrine was deep fires into the rear echelons of the advancing enemy army, a concept employed with devastating effect against the defending Germans and Japanese in World War II, the North Koreans and Chinese in the Korean War, and the North Vietnamese in the 1960s and 1970s. Any reader of the successive army Field Manuals 100-5 from 1940 through the 1990s would be struck by the broad continuity of thought, the identical phrasing of many salient passages, and the unremitting focus on force-on-force warfare. With at least one of the pillars intact, the army limped into the new millennium.

Losing Clarity: The Rise of Full-Spectrum Operations

In 2001 the army renumbered the Field Manual 100-5, "Operations," series to become the 3-0 series, synchronizing and subordinating its doctrine to the new joint doctrine emerging from the Joint Chiefs of Staff. The new joint doctrine, which has become known as "Full-Spectrum Operations," completely abandons the army's heritage of firepower-based attrition warfare, discussed previously, and postulates a battle space dominated not just by lethal fires but also by nonlethal effect and information operations. Synchronizing these effects would, it was hoped in 2001, produce rapid and relatively bloodless victories. Joint Publication 3-0, "Joint Operations," now views military operations in a full-spectrum continuum from "military engagement, security cooperation, and deterrence activities to crisis response and limited contingency operations and, if necessary, major operations and campaigns."[6] Major operations and campaigns, we are told, involve large-scale combat, placing the nation in a wartime state. The general goal is to prevail against the enemy as quickly as possible, conclude hostilities, and establish conditions favorable to the host nation, the United States, and our multinational partners. Stability operations are often required.[7] Operations may be terminated by either imposed or negotiated settlements, and termination must be considered from the outset of planning. Remarkably, General Schwarzkopf's conference with the Iraqi army after Desert Storm is the illustrated example.[8] Offensive and defensive operations (combined!) merit a single short paragraph, almost as if the difference between them were of no great significance to the overall campaign.[9]

Joint Publication 3-0 tells us that major operation and campaign plans must feature an appropriate balance between offensive and defensive and stability operations in all phases. However, stability operations are depicted as the most important in all the notional phases.[10] "Stability operations typically begin with significant military involvement to include some combat, then move increasingly toward enabling civil authority as the threat wanes and civil infrastructures are reestablished. As progress is made, military forces will increase their focus on supporting the efforts of host nation authorities . . ."[11]

In its essence, the joint doctrine denies meaningful differences between offensive and defensive wars or in the nature of the stability operations consequent to either. Indeed, the new doctrine contains no doctrinal terms equivalent or akin to "offensive war," "defensive war," "occupation," or "liberation," even though these overarching ideas have driven American military thinking from the earliest days of the republic. It also fails to assign specific responsibilities to the various component commanders—land, air, and maritime—for the various aspects of the war. The overall impression is that all matters are situation dependent, all options are open, and joint combatant commanders, no matter what their service, and political authorities, no matter what their familiarity with military affairs, will make the right decisions at the appropriate times. The doctrine is so vague as to defy critical analysis and empirical validation.

Current army capstone doctrine unsurprisingly mimics joint full-spectrum doctrine from a "landpower" perspective. Field Manual 1, "The Army," echoes that "Army forces execute a simultaneous and continuous combination of offensive, defensive, and stability and reconstruction operations as part of integrated joint, interagency, and multinational teams."[12] Stability operations are elevated to be coequal with offense and defense. The 2008 rewrite of army Field Manual 3-0, "Operations," which revises and in minor respects improves the 2001 version which informed our wars in Iraq and Afghanistan, predictably parrots the same fruit salad theme: offense, defense, and stability efforts are intermixed in all operations, their relative weights being decided by the mission at hand.[13]

In its latest edition, offense is discussed only at the operational and tactical levels with respect to defeating enemy combat forces. The stated purpose of offensive operations is no longer a strategic aim—the "destruction of the enemy's armed forces"—as it was in the World War II doctrine. Instead, the doctrine suggests a menu of enabling lesser tasks: dislocating, isolating, or disrupting enemy forces; seizing key terrain; depriving the enemy of resources; gaining information; deceiving and diverting the enemy; and creating a secure environment for stability operations. The types of offensive operations are solely tactical: movement to contact, attack, exploitation, and pursuit.[14] Defensive operations similarly are discussed only at the tactical level of war.

The Rediscovery of Stability Operations

Stability operations are a grab bag of all overseas actions other than offense or defense. "Stability operations encompass various military missions, tasks, and activities conducted outside the United States in coordination with other instruments of national power to maintain or reestablish a safe and secure environment, provide essential government services, emergency infrastructure reconstruction, and humanitarian relief (JP 3-0). Stability operations can be conducted in support of a host-nation or interim government or as part of an occupation when no government exists."[15] No differentiation is made in FM 3-0 between stability operations requirements for offensive vice defensive wars, occupation vice liberation, or military government vice civil affairs, even though these distinctions had been thought critical in previous American wars. Indeed, words like "occupation," "liberation," "military government," and "martial law" are no longer part of the army's list of doctrinal terms.[16]

The inclusion of security operations in the new doctrinal formulation should alarm any student of the history of U.S. doctrine. Only once before in our history, during our failed effort in Vietnam, has the army embraced stability operations as coequal with offense and defense.

The term "stability operations" was first coined in 1964 by army chief of staff Harold K. Johnson to associate a collection of activities—counterinsurgency, nation-building, security assistance—that was hoped to allow friendly Third-World governments to withstand Communist aggression. The phrase was dropped from army doctrine in 1974, and the various ideas it represented were demoted to second-tier military missions.[17] The failure of nation-building in Vietnam caused the army to abandon stability operations in favor of its institutional preference, force-on-force warfare.

> In contrast to the 1968 edition of FM 100-5, which had stated that "the fundamental purpose of U.S. military forces is to preserve, restore, or create an environment of order and stability within which the instrumentalities of government can function effectively under a code of laws," the very first sentence of the 1976 manual declared unequivocally that the Army's primary objective is to *win the land battle*—to fight and win battles, large or small, against whatever foe, wherever we may be sent to war . . . Gone were all references to counter-insurgency, nation-building, civil affairs, and psychological operations, replaced by a single-minded emphasis on the conduct of conventional combat operations in a major war.[18]

Today, similar nation-building missions in Iraq and Afghanistan bring the army back to the mid-1960s formulation; "stability operations" once again stand center stage as our ill-considered installation of flawed overseas regimes compels us to "stabilize" what we've created.

Flawed Taxonomy

Field Manual 3-0, like Joint Publication 3-0, arrays conflicts by size and intensity in a spectrum that ranges from stable peace to unstable peace to insurgency to general war. Oddly, only insurgency is recognized as a distinct form of warfare.[19]

The view that wars do not fundamentally differ in "kind" but rather vary mainly in their intensity is ascribed to Clausewitz.[20] Clausewitz, too, envisioned war as a contest of wills between sovereigns, played out at different levels of violence, with combinations of offensive and defensive battles. As an overarching philosophical statement, this construct is unchallengeable. But at a practical level, where a military professional must use observations of past wars to make positive, predictive models of current and future wars, the Clausewitzian construct adopted in our Full-Spectrum doctrine is meaningless.

To practitioners, taxonomies matter. Sorting objects and phenomena into meaningful categories is the foundation of all science and analysis. For instance, modern medical understanding would not be possible if biologists failed to divide life forms into plants and animals, animals further into vertebrates and invertebrates, vertebrates into mammals and reptiles, and so forth. An observation that some life forms are large and others small is not nearly as important as recognizing differences in fundamental type or quality. Sorting life forms primarily by size, not by kind, is something that no serious scientist would do. Yet both our joint and army doctrines sort warfare primarily by size and intensity, as if these characteristics were the defining differences in wars.

In the new doctrine, operations within wars are sorted into offense, defense, and stability, yet wars themselves are never discussed as being offensive, defensive, or stabilizing, a denial of taxonomic distinction. Failure to differentiate between the different kinds of wars predictably leads to an inability to properly analyze and recognize the patterns in wars, which leads to haphazard results in predicting the requirements for concluding wars victoriously.

Rediscovering Counterinsurgency

The strategic failings of the joint and army "operations" doctrines carry forward into the recently published Field Manual 3-24, "Counterinsurgency," which some in the press publicly extolled as a breakthrough document.[21] The manual begins by correctly stating that "political power is the central issue in insurgencies and counterinsurgencies,"[22] then suggests that insurgencies form to overthrow unpopular indigenous governments or, alternatively, as resistance movements seeking to expel or overthrow foreign or occupation governments.[23] Unfortunately, this strategic dichotomy—insurgents as aggressors or insurgents as defenders—is not retained in the remainder of the manual. Counterinsurgency becomes another "Full-Spectrum Operation" inherently neither offensive nor defensive in nature.[24]

The central discussion contained in chapter 5, "Executing Counterinsurgency Operations," of FM 3-24, refers constantly to a host nation—shortened to the acronym HN—implying that we are invited in by an indigenous government to assist in its defense. The manual is remarkable in that it fails to appreciate that the U.S. Army might in some cases be the invading and occupying foreign power, which an indigenous population is resisting by guerilla warfare.

Indeed, the greatest flaw of FM 3-24 is its failure to fully appreciate the difference between insurgencies aimed at defeating foreign invaders and insurgencies aimed at overthrowing unpopular homegrown regimes. Consequently, our counterinsurgency operations undertaken as part of an offensive war—for example, Iraq—are confused with the defensive measures that an established government takes against local malcontents in, say, British Malaya or

Northern Ireland. Of course the army would try to use the minimum force necessary against insurgent American citizens, or even citizens of a genuine host nation that asked us to assist in its counterinsurgency effort, and of course one would use mainly political and economic levers to try to ameliorate popular discontent among one's own people. But history shows that imposing rule over a hostile foreign population after an invasion is a much bloodier affair. The invader should expect a flood of nationalism—or nativism— to blossom into insurgency. The Philippines, Transvaal, Tibet, Vietnam, Afghanistan, and Iraq, all discussed previously, are but a few of the numerous historical examples.

Nor should the attacker be surprised if the chaos and passions unleashed by his invasion take the invaded country in a starkly unfavorable direction. For example, Japan's 1937 invasion of China unintentionally assisted Mao in his eventual victory over the Kuomintang. Further, our invasion of Iraq and ouster of the Saddam regime unleashed insurgencies by both Sunni and Shiite religious fundamentalists that we did not foresee or desire. Preventing and eliminating post-invasion insurgencies through security operations and enlightened military governance must be a prime focus of any offensive military doctrine, a lesson learned in the army's past but conspicuously absent from the new counterinsurgency field manual.

Field Manual 3-24 tries to codify our current "stability" operations in Iraq into doctrine. Perhaps, given events on the ground, that effort is premature. FM 3-24 explicitly rejects the army's hard-war heritage, stating that force should be "measured" and that "the more force you use the less effective it is"—a recognition that violence initially begets only more violence but a denial that wars by default become wars of attrition that will generally last until the defender loses the demographic capacity to further resist. Sadly, wars, once joined, tend to escalate and run their demographic course before the blessings of peace can once again be properly appreciated. History teaches that the savagery of war and the bounties of peace do not coexist.

Field Manual 3-24's counterinsurgency "operational design" is long-term programs aimed at winning popular support through better security, more jobs, improved utility service, good governance, and other mainly nonlethal lines of operation. The hope is that we can isolate and eventually defeat insurgent bands without resorting to full-fledged, bloody counterinsurgency war.

Much emphasis is placed on raising indigenous forces during the time of war, to which we will, as the forces become capable, hand over the bulk of security responsibilities. Many of these approaches have been tried before—for instance, in Revolutionary America by the British, by the U.S. Army in the Philippines and Vietnam—and have generally proved ineffective, as regrettably now may be the case in Iraq and Afghanistan.

On a positive note, in the key chapters, Clausewitzian center of gravity analysis is never mentioned. Instead, the text requires a "systems approach" to understanding the enemy, still another concept crafted to attempt to avoid attrition warfare.

The Operational Focus: War Without End?

However, the greatest failing of the Full-Spectrum Operations doctrine, both joint and army, is the focus on selected operations and tactics, not on war and strategy holistically. War is never defined or categorized. Victory is never defined. Indeed, the very idea of war as a distinct definable entity with a beginning and an end—that is, the constitutional formula of a declaration of war at the beginning and a peace treaty at the end—is never discussed. Instead, the doctrine invokes the vague notions of "persistent conflict" and "the long war." Conflicts become an endless stream of "operations." Gone is the episodic nature of war envisioned by the Founding Fathers; instead, the military pictures itself in constant conflict on the various and ambiguous frontiers of our new Pax Americana. Missing in the new doctrine is a notion of final victory against which to measure the effectiveness of our strategy or combinations of operations. We now can scarcely measure if we are winning, or how fast we are winning, or even define what winning means. The military just continues to conduct operations in the hope that good will come of it.

This lack of strategic context and the notion that war is not a boundable phenomenon that can be scientifically analyzed makes Full-Spectrum Operations both irrefutable and unverifiable. As such the new operations focus is a dangerous doctrinal excursion, offering neither the military nor our civilian leadership any confidence in our ability to win wars within the time and resource constraints normally applied by democratic governments and their electorates. The military leadership may embrace an open-ended operations framework, but the civilians for whom we work think in terms of wars that have definable beginnings, ends, and costs. The failure of our doctrine to address their concerns diminishes its relevance and applicability.

The military should categorize and define wars in ways that laymen can understand. For instance, sandwiched between two Clausewitz quotations, Field Manual 3-0 does provide a useful and concise definition of landpower: "the ability—by threat, force, or occupation—to promptly gain, sustain, and exploit control over land, resources, and people . . . impose the Nation's will on adversaries . . . establish and maintain a stable environment that sets the conditions for political and economic development . . . restore infrastructure and reestablish basic civil services . . ."[25] Bravo! For a brief, lucid instant, Field Manual 3-0 hints at a definition of "offensive war"—my term, not the army's—as a strategic role that only the army can fill, yet it merits only a passing mention in our capstone "Full-Spectrum Operations" doctrine. Nor does FM 3-0 follow up the definition with a detailed discussion of the various lines of effort and campaigns that would be necessary to obtain the desired offensive "victory." The necessary linkage of war to national strategy, of campaigns to war, of operations to campaigns, of tactics to operations, so vital to the successful prosecution of our craft, lies unexamined in our capstone doctrines.

Why We Need Doctrine for Offensive Wars

The absence of strategic offensive military doctrine leaves our senior military leaders ill-prepared to advise the president on offensive warfare. We lost our way because we rejected our attrition war heritage, our Cold War absorption in defensive thinking, the navy's and air force's majority influence on joint doctrine, the army's institutional preference for force-on-force warfare and aversion to nation-building, and our country's exaggerated self-image as a welcomed "liberator" around the world. Clearly, historically validated doctrine on how to wage wars of occupation was not available to President Bush as he decided to invade Iraq in 2002, nor is such doctrine available now in the Full-Spectrum construct.

The U.S. Army, the only armed force able to impose new governance on a significant foreign population, owes the president and the nation a clear explanation of the method, costs, and risks associated with this type of war. The doctrinal prescription need not be speculative; the United States Army has a rich history of successful offensive warfare. We can also examine the successes and failures of foreign nations. All the necessary data is available in the historical record: the strategic, operational, tactical, and even subtactical experience. Empirical doctrine could be written.

By now the outline of America's rediscovered offensive doctrine should be no secret to the reader. Landpower, in the form of the U.S. Army, is the main effort in offensive wars of occupation. These wars demand a prolonged and substantial new national strategic commitment, over and above the force structure devoted to peacetime engagement and routine national defense, and require mobilization of additional national resources. To achieve the compliance of the population to American rule, grievous casualties will likely have to be inflicted on the targeted nation's population, well beyond the defeat of their immediately existing military. The United States will have to assume responsibility for supervising all governmental functions in the occupied country, including ensuring the security and well-being of its citizenry. Over time, we will have to transition governmental authority to the occupied population and build a new indigenous government supportive of our vital interests. Getting our defeated former enemy to ally with us willingly and permanently is the capstone achievement defining victory. Over time, lasting democratic institutions may perhaps develop with American assistance. Finally, failure to develop acceptable governance could unleash revolutionary forces more threatening to our strategic interests than our enemy of the moment.

If this narrative makes offensive wars of occupation less palatable to the nation, so be it. The role of doctrine is not to make military action seem falsely attractive but to explain the sometimes awful requirements for victory. Suitable military and policy options other than offensive wars of occupation are generally available to America. Invasion of another country for purposes of establishing a new government must be seen as the extreme strategic commitment that it will predictably become.

These may be messages that many would rather not hear, but it would be dishonest to state anything otherwise. Artful constructs that make war seem a convenient policy option reflect self-deception, not careful study of the reality of warfare. War, even when necessary, is a truly awful thing. It should never be entered into lightly. Its true cost should not be wished away. War should be the last alternative, forced rather than chosen. But when forced, there must be understanding that the violence we unleash will test the limits of conscience. Our forces will not apologize for the necessary killing of the enemy's soldiers or population. We will be generous only after victory is achieved. Once we again understand these truths, we will wage war less often, but we will terminate wars more successfully.

Chapter 10

Organizing for Military Governance

lthough there is a growing realization in the U.S. defense and international affairs community that a period of military governance must be the
final phase of offensive wars, there is little agreement about the specifics.
To date, clear responsibility for postwar governance has yet to be assigned to any
specific department of the U.S. government or within the Department of Defense.
Consequently, the U.S. government has developed no meaningful doctrine or force
structure that would prevent a new Iraq-like debacle should the nation ever in the
future wage offensive war.

As we shall see below, there are plenty of ideas about the creation of new
departments, agencies, and commands to address the problem, none of which has
acquired or is likely to acquire the institutional momentum to bear results. For a
variety of very practical reasons, only the army can perform military governance
in the national interest. Based on its own assessment of its institutional interests,
however, the army's senior leadership has been disinclined to embrace the military
governance mission or pledge meaningful resources to its accomplishment. I argue
that the army should not be allowed to evade the governance mission and should
be made to train and organize forces for that purpose, even if it means trading
away units specialized for force-on-force battle.

Our point of entry into the current discussion must be the highly influential works of Thomas P. M. Barnett, a U.S. Naval War College professor
and the Assistant for Strategic Futures, Office of Force Transformation, in
the Office of the Secretary of Defense from 2001 to 2003. Using that office
as his platform, Barnett spread around Washington, D.C., the neoconservative
ideas that would later be published in his 2004 *New York Times* bestseller, *The
Pentagon's New Map*. Barnett argued that the United States, as the world's sole
superpower, must compel the unconnected, lawless nations of the "Gap" to join
the "Functioning Core" of connected, rule-abiding nations, using military force

if necessary.[1] Assessing the merit of Barnett's neoconservative call for a new era of offensive wars is a subject for later chapters. Of more interest here is Barnett's lucid observation that, although the military currently has the ability to destroy "Gap" regimes, it does not have the ability to either effectively occupy these defeated nations or replace the old regimes with better governments. Barnett urges the creation of a second military to "wage peace" and bring conquered nations into the fold of the rule-following Core. Barnett names this new military "the System Administrator force."[2] He is never clear about how this System Administrator force would be constituted, at times implying that the lead would be with international partners, at times the private sector, at times the U.S. Marine Corps.[3]

Those expecting greater clarity in his follow-on book, *Blueprint for Action*, published the following year, would be disappointed. Barnett laments the institutional bias of the army against the nation-building mission. "But of course it is the Army that is most resistant to the notion of the SysAdmin force, fearing that it will lose its warfighting ethos and be turned into a giant peacekeeping force."[4] The Defense Department is similarly scolded: "So long as the SysAdmin's main assets remain buried and poorly prioritized inside the Defense Department, it will always be the unfunded mandate it is today."[5] Barnett observes that:

> [the Defense Department] can project power around the planet with great agility, conducting wars of ever-shorter duration while fielding a much smaller force that suffers far lower casualty rates. But our ability to field an effective nation-building response, which naturally remains manpower-intensive, has not kept pace, primarily because the Pentagon has long refused to invest in such capabilities. . . . [There's a] new rule the Pentagon needs to adjust to: the shorter the warfighting, the longer the peacekeeping. The smaller the war force, the larger the peace force. The easier the war, the tougher the peace.[6]

Convinced that the SysAdmin force will never find a proper home in the Defense Department, Barnett reluctantly calls for the creation of a new government department to perform the nation-building function in order to bridge the gap between war (Defense) and peace (State).[7] Later, Barnett estimates that the SysAdmin force would be one-half military, one-quarter civilian police, and one-quarter government workers, and that the American contribution to all three aspects should "hover in the 10–20 percent range." [8] It is hard to see, given Barnett's formula, how any U.S. government agency, foreign nation, or international body would assume the institutional responsibility for making the SysAdmin force happen.

While Barnett was famously touting his SysAdmin concept in 2004, the National Defense University, an arm of the Department of Defense, also recognized, albeit more obscurely, the need for a dedicated force for post-conflict operations and recommended a solution within DOD. The capstone recommendation was for the creation of two stability and reconstruction joint commands (S&R JCOMs), one on active duty and the other in the reserves, designed to provide capabilities for post-conflict nation-building or pre-conflict assistance to friendly governments. These new commands would be "modular" and "scalable," "tailorable to mission requirements," and commanded by flag officers at the one- to three-star level.

The main capability of the S&R JCOMs would reside in its four Joint S&R Groups, each of which would consist of a military police battalion, a civil affairs battalion, an engineer battalion, a medical battalion, and a psychological operations battalion. The S&R JCOM could also be assigned combat units—armor, infantry, artillery, and attack helicopters—as necessary. Depending on the mission, a S&R JCOM would field between 11,300 to 18,200 personnel from the various uniformed services.[9]

Notably, all the subunits of the new S&R JCOMs currently exist in the force structure, either in the reserve or active component, so the establishment of the new division and brigade-level headquarters is the only real initiative here. Never in the discussion is there a recognition that the S&R JCOMs or groups might actually be responsible for the military governance of occupied nations or provinces. The text implies that the new organizations would perform tasks supportive of governance, but actual governmental authority would reside in some other unnamed entity. By recommending creation of the new headquarters in the joint arena, rather than in a specific service such as the army, the study's authors have further avoided assigning clear responsibility for mission success.

The army further distanced itself from service responsibility in September 2006, when the Association of the United States Army, the army's unofficial think tank and lobbying organization, published its thirteen-page study of the issue entitled "SysAdmin: Toward Barnett's Stabilization and Reconstruction Force."[10] The brief foreword, by former army chief of staff Gordon Sullivan, recognizes the need for "new emphasis [on] Stability, Security, Transition and Reconstruction (SSTR) operations" and gives credit to Barnett for his contribution but calls for the creation of an independent joint command for SSTR within the Department of Defense.[11]

The text later refers to the new command as "USPEACECOM." The specific details about what would constitute USPEACECOM are stated to be "outside of the scope" of the paper. Also up in the air is the contribution the army would be willing to make to the new command.[12] The document can only be interpreted as a statement of the army's institutional disinterest in being the solution to a problem it recognizes as urgent. Never does the study's author argue that the army should be the logical proponent for the SSTR mission.

Stopping whatever progress may have been ongoing in the Department of Defense, National Security Presidential Directive/NSPD-44, Management of Interagency Efforts Concerning Reconstruction and Stabilization, signed by President Bush on December 7, 2005, assigned to the State Department responsibility "to prepare, plan for, and conduct stabilization and reconstruction activities." The secretary of state, and the subordinate coordinator for reconstruction and stabilization, are to be the "focal point" and "lead" for the interagency effort. Although coordination with the secretary of defense is directed, the secretary of state is answerable to the president for the postwar reconstruction and stabilization mission.

The root problem with NSPD-44 is the State Department's inability to put sufficient numbers of people into an occupied nation to perform the mission for which it has been made responsible. The State Department's mantra that the army has more people assigned to musical bands than the State Department has in its foreign service attests to its institutional inability to meet the scope of the problem.[13] Indeed, in February 2007 the State Department had to acknowledge that it could not fill about half of the three hundred slots it had signed up for on Iraqi PRTs and requested that the Department of Defense fill them instead.[14] In 2008 a senior advisor in the State Department characterized the interagency effort as "ones-ies and twos-ies" rather than the hundreds and thousands of dedicated personnel actually required.

The State Department's solution appears to be the proposed Civilian Reserve Corps, which Congress authorized in 2008. In charge of this interagency effort, the State Department's coordinator for reconstruction and stabilization has developed a three-tier system for producing the needed deployable workforce. The Active Response Corps (ARC) would consist of approximately a hundred State Department, USAID, and other federal interagency full-time employees ready to deploy on forty-eight-hours' notice and stay deployed for up to six months. The second tier, the Standby Reserve Corps (SRC), would number another five hundred to two thousand federal employees who, within thirty to sixty days of a crisis being determined, could take leave from their current positions to deploy to the overseas area. The thinking is that this cadre would add another two hundred to five hundred stability personnel overseas, given a projected 25 percent deployment rate. The final tier, the Civilian Reserve Corps (CRC), would consist of hundreds of volunteers from state and local governments, the private sector, and nongovernmental organizations who would make a four-year commitment to serve overseas for up to one year if needed. Again assuming a 25 percent deployability rate, this tier could eventually produce several hundred more overseas personnel for the State Department.[15]

Although laudable, generating several hundred interagency experts for overseas stability operations is different from actually fielding the thousands of people in the right organizations, properly trained and equipped, who will actually be required for postwar governance. At best, the ARC, SRC, and CRC combined could only augment what will still be a largely military effort.

The State Department may want the nation-building mission but has too few assets to perform it, with most of the needed deployable force structure and institutional capacity being in the army. A senior advisor to the coordinator for reconstruction and stabilization recently confessed that the army knows how to train, plan, organize, and resource for these sorts of overseas missions far better than does the State Department. The expeditionary expertise and capability is in the military, not the interagency. Creating a parallel "quasi-military" structure in the State Department would only be duplicative and wasteful.

In many ways, the current situation is a replay of a World War II debate when Secretary of State Cordell Hull wanted the Germany reconstruction mission. President Roosevelt initially wanted to give it to him for a variety of reasons, but as a practical matter the president had no real option but to give the mission to the army. Only the army could field the necessary units in the necessary numbers with the necessary resources at the appropriate times.

The NSPD-44 framework remains mired in bureaucratic infighting between State and Defense—State wants the mission but can't do it; Defense doesn't want the mission but also doesn't want to put any of its force structure under State control. One very senior army officer stated in 2008 that the nation's shortfall in institutional stabilization capability was a "most critical national security concern," but any proposal to move the mission from the State Department back to the army would have to await the 2009 inauguration of a new administration.

Although the army leadership evades the national governance role, its new doctrine does invest local governance responsibilities in lower-level combat unit commanders by assigning them areas of operation and making them responsible for conducting stability tasks along multiple lines of operation that include developing and providing "governance, essential services, host nation security forces [including police], and economic development."[16]

In Iraq many lower-level combat unit commanders have become the de facto or shadow governors and mayors within their assigned areas of operation, by default responsible for a broad range of responsibilities for which they, and their subordinates, are professionally unprepared. Official army histories have consistently remarked that armor and infantry officers and enlisted men who have spent their careers training for force-on-force combat simply lack the kinds of political and technical skills needed to run a municipality or province, especially among a foreign population.[17] Even with the addition of a small G-9 or S-9 civil-military operations (CMO) staff section or a few CMO teams to assist civilian liaison, the overwhelming character of the army's combat units in terms of manning, equipping, training, and personality makes them a poor fit for the civilian governance mission. Only exceptional combat units with uniquely gifted commanders can overcome their war-fighting biases to succeed as municipal or regional governors.[18]

The army policy of rotating combat brigades in and out of their areas each year further exacerbates the governance or governance support mission by continually breaking whatever working relationships have developed between the local population and the combat forces in an area of operations. The combat units tend to voice similar frustrations: they are deployed on a mission they are not designed for or fully prepared to conduct; they are operating in countries and environments they do not adequately understand; and they receive little or no help from the State Department or other federal agencies.

Given the recognition that much of the responsibility and capability for postwar reconstruction must come from the Department of Defense, one would expect that the recently approved end-strength increases for the U.S. Army and U.S. Marine Corps would provide the personnel spaces the force structure required to man the new "stability" headquarters and units proposed by the various experts regardless of who's in charge.

But nothing of the sort has happened. By shoving the responsibility off on the joint, interagency, or international community—a nebulous external amalgam—all U.S. government bureaucracies have in essence shielded themselves from the mission that none of them really wants. The army will use its sixty-five thousand new spaces to build six more brigade combat teams and support brigades, all designed for force-on-force combat, rather than invest in military governance capability. Caught in a vicious Catch-22, the army feels that it must expand its baseline of combat brigades committed to rotations in and out of what has seemed to be a never-ending war in Iraq, which at best appears only to be shifting to Afghanistan, rather than develop a sound approach to military governance that might make that scale of rotations unnecessary.

For their part, the marines are building more marine infantry regiments and supporting arms, which is completely appropriate for their mission within the Department of Defense but offers nothing to the stabilization and reconstruction mission. The air force and navy are actually cutting their manpower as a bill payer for their high-dollar weapons-procurement programs, so they can hardly be called upon to sign up for new joint billets in newly created joint stability commands.

Therefore, despite the largest defense buildup in the last two decades, there is no plan to build the kind of postwar reconstruction capability that all agree is necessary. All we have are prosaic words, empty promises, and behind-the-scenes evasion of responsibility.

Creating an Army Military Governance Division

Any proposal to make any organization other than the army responsible for postwar governance and reconstruction is doomed to fail. The army is the logical and practical choice for many reasons.

First, the army is the nation's military force dedicated to major land campaigns and will always provide the backbone of the land effort, even if the marines participate. Governance and reconstruction happen on land. If they happened at sea or in the air, then the navy or air force would be the logical service choices to execute the necessary post-conflict missions, but that is not the case.

Second, the bulk of the functional units needed to support governance and reconstruction missions (military police, engineers, medical units, civil affairs, ground transportation, and supply) is part of the army force structure. These units are primarily designed to support ground combat operations, but they can also support governance and reconstruction missions if required. Indeed, as a practical matter, there is often a need to prioritize available support assets between conflicting combat and population support missions. The necessary allocation and prioritization function is best accomplished when a single headquarters is responsible for both the land combat and post-combat mission.

Third, the need for military governance, security, population control, and emergency humanitarian assistance at any given location occurs within hours (not days or weeks) of successful combat operations. To use Eisenhower's metaphor, behind the advancing fringe of combat troops there is an immediate need to lay down a carpet of governance, security, and emergency assistance that will be the beginning of the reconstruction. This close coordination with the attacking ground combat units can best be accomplished by other army units specialized and trained for the governance and security task. Creating an interservice or interagency seam immediately behind the combat fringe is simply unworkable from a command and control standpoint.

Fourth, occupation of a foreign country always invites the risk of insurgent warfare. Counterinsurgency strategy requires close coordination between the counterguerilla military effort and the general governance of the population. Keeping the military effort and the governance effort unified in a single command until the risk of insurgency has passed is therefore a military necessity in offensive war.

Fifth, any scheme that relies heavily on sufficient numbers of interagency experts and representatives actually deploying overseas in a timely fashion flies in the face of the demonstrated inability of the interagency to mobilize the needed capability, especially in the absence of conscription. The army's World War II military governance model worked best. The CORDS interagency structure was amply resourced, but it came too late in the Vietnam War to make a difference. Interagency contributions in Afghanistan and Iraq have been embarrassingly inadequate for a government desiring to pursue overseas military adventures.

Finally, and perhaps most importantly, unity of command and unity of effort are required to win wars. A single, overall commander needs to be made responsible for seeing the offensive war effort through to successful conclusion. Because all offensive wars are land campaigns, the army commander is the only logical choice to command the conventional fight, the initial occupation, and the military governance phases of the mission. History offers no alternative model for success.

The heart of the army's governance capability would be an entirely new organization for today's army drawn from successful practices of the past: a military governance division, a unit of perhaps ten thousand uniformed governance specialists, many of which would be reservists or former civilians specifically brought on to active duty to perform occupation functions.[19] Of this number, perhaps three thousand would be in the division headquarters, assigned the mission of running the national government of the occupied country, and the remaining seven thousand would be organized into brigade, battalion, and company units designed to govern provinces, cities, and smaller population groupings, respectively. The various headquarters staffs would be organized along governmental lines, not by traditional military functions. For example, a military governance division staff would not be organized along the G-staff model: G-1, personnel; G-2, intelligence; G-3, operations, et cetera. It would be organized along the lines of the departments of the executive branch of a national government: treasury, justice, state, defense, agriculture, energy, education, public health, et cetera, and have other staff sections responsible for the national legislature and courts.

Similarly, governance brigade headquarters would be organized along the lines of state governments or major cities, representing most of the functions found at the national level, minus currency policy, foreign relations, and other functions performed solely at the highest level. Governance battalions would mirror county and local governments: police, fire, public works, identification cards and records, lower courts, et cetera. Governance companies would comprise teams capable of supervising governance at the lowest levels: boroughs, neighborhoods, and towns.

Importantly, the military governance division would not be staffed to perform all governance functions but to dictate policy to and supervise the government workers of the occupied nation.[20] Use of the currently existing government workforce would be key to successful occupation, there simply being no practical way for the United States to directly govern a foreign population. Neither would the division have the inherent capability of providing services to the population other than governance. Service workers in the occupied nation—for example, hospital staffs, policemen, water and sewerage workers, firemen, and teachers— would still execute their professional responsibilities under the direction of the military government. Only by exception would the military government division require U.S. military units or other coalition assets to provide public services to the occupied population. As a practical matter, most all of the work would have to be done by the locals anyway because the ability of the occupying military force to provide widespread and enduring assistance to the population always would be limited. Nor would it be desirable to make the occupied population dependent on the occupier's services when it should instead be increasing its own capabilities to provide those services.

The clear exceptions would be policing and criminal justice, which would become the direct responsibility of the occupation army as it establishes its monopoly of violence in the conquered land. Military police and constabulary battalions, working under the direction of the military governance division, would have to provide the necessary policing functions while a new local police force would be rapidly stood up out of the vetted personnel from the police force of the overthrown regime. Military officers would have to prosecute and judge criminal cases in summary courts while the new civilian court system was being established.

In peacetime a military governance division would be mainly a cadre organization of about a few hundred soldiers, a mixture of active duty and reservists. Perhaps two would exist in the force structure, it being improbable that the nation would launch offensive wars against more than two nations at the same time. The division's principal mission would be to participate in joint planning and exercises revolving around offensive war scenarios. Importantly, it would also maintain and update manning documents detailing—insofar as possible—how the organization would be filled out by army, other armed service, interagency, and civilian experts in case it were mobilized for occupation duty. Upon mobilization, the division would reorganize to align its staff in accordance with the governmental structure of the targeted nation. For instance, if the United States was invading a country with an oil ministry and seventeen provincial governments, then a staff section for oil would be created within the division headquarters, and the necessary brigade headquarters would be constituted to align with the seventeen provinces. Religious affairs sections may also be needed in areas where religious leaders wield political influence. Within a matter of months, thousands of people must be brought into the military governance division, be organized and trained for their wartime tasks, be allowed time to coordinate and refine their plans, and be deployed overseas for employment.

Getting access to the right people would be crucial. Critically, the army must realize that nonstandard personnel practices would be required. Very few of the people with the skills necessary to establish and supervise an effective civil government will be found in the active-duty force. Much of the staff would likely come from academia, private industry, various branches of the national government, and state and local agencies. Some may be foreigners with specialized knowledge. Civilian translators and people with local expertise must be contacted and hired immediately. In cases where the needed people are not already employed in the federal government or in the reserves, direct accession and commissioning of the necessary people into the army might be the best option. Turning college professors into colonels and cabbies into corporals, however unusual elsewhere in the army, might be the norm here. Most, if not all, of the military governance division should be uniformed military, with only a small number of other federal employees attached, as was the practice during and after World War II.

The deputy commander for military governance would exercise unusual authority for a military officer and must possess skills not normally developed in the narrow confines of the mainstream military. He must also enjoy the full confidence of the Defense Department, State Department, and eventually the governed population. He must, therefore, be carefully selected.

The military governance division, in consultation with the State Department and other governmental agencies, should try to plan the end-state character of the new national governance even before the war begins. However, the review of the historical data shows that events generally overtake plans, in part because plans are never made or agreed on in timely fashion, and partially because the local population always gets a vote, especially when the overall goal is increased democracy and popular participation in the institutions of the nation. Perhaps the best that prewar planning can do is to inform the occupiers of the traps that await them as the various elements of the occupied nation jockey for power in the new regime. Every nation is different, and there is simply no cookie-cutter formula for imposing new governance that works everywhere.

Neither is it true that all is possible everywhere at all times, even given the best possible organization, resources, and intentions. Some problems in some nations simply defy reasonable solution by military occupation. There is a clear need for the military governance division to establish the limits of what is achievable and what is not, determine the risks associated with the occupation, and come up with center-line forecasts of the time and effort involved in obtaining the desired political end state. This governance estimate should be a critical component in determining whether the proposal for offensive war should be approved and also inform the operations and rules of engagement that will be used by the combat forces if invasion is ordered.

Employment of the division during wartime would be along the lines of the World War II model for Germany. The division commander would become the land component commander's deputy commander for governance. The military governance brigades and battalions, based on their predetermined alignments with existing political subdivisions, would roll out a carpet of governance immediately behind the advancing combat fringe in support of and in coordination with the ground combat commanders and their G-9 (civil-military operations) staffs. At each point on the terrain, combat and military governance units would be intermixed for a period of days, meaning that the initial actions of the military governance detachments must be under the direction of combat commanders. But as the combat units move on, the military governance units would stay behind in their designated localities to continue the governance function. To the civilians on the ground, the process would be made predictable: combat, issuance of the proclamation declaring occupation, necessary clearance by the combat forces, and immediate reestablishment of local government supervised by American soldiers and enforced by military police and American military courts. Eventually, U.S. forces would reach and occupy the national capital, and the division headquarters would immediately assume responsibility for the functioning of the national government and its ministries. Over a period of a few years, those ministries and other agencies of government would be re-formed along democratic lines in accordance with local realities and American national desires. The conquered population should be allowed to reestablish its own sovereign government only when it proves itself capable and trustworthy.

Aside from the period of initial contact during the combat phase, tactical combat commanders should not have operational authority over military governance units, nor should military governance units ever command combat forces. The historical record indicates that these two divergent responsibilities should be kept as separate as possible, as it is impractical to govern people you're fighting, or fight people you're governing.[21]

The military governance unit commanders operate from the presumption that enhanced law enforcement using police (military and/or local) and courts (summary and/or local) will be able to maintain law and order. Constabulary units may be constituted and trained for riot control and civil disturbance missions, but they should conduct themselves as law enforcement agencies or national guardsmen would handle similar situations in America. Failure of citizens of the occupied nation to comply with the laws of the occupation authority should be handled as criminal matters, and the accused should be allowed the most liberal due process reasonable under the circumstances.

Should insurgency develop within the occupied country, the overall commander—that is, the land component commander or joint force commander—must have combat units available under his command to address that situation. There arises almost a good cop/bad cop choice for the occupied population: either comply with the just requirements of the military governance units or face renewed fighting and destruction at the hands of combat forces. Although there undoubtedly would be a contest of wills between the occupier and the occupied, the goal would be to resolve the conflict in our favor with the least amount of postwar violence possible.

Establishing New Indigenous Governance

Based on the historical review, restoring full sovereignty and a new "normalcy" to the conquered nation must be seen as a labor of years. From the beginning of the occupation, governing through vetted local officials is desirable and necessary. But restoring full local authority at the various levels of government and full sovereignty at the national level must be a measured process. At each stage the occupied population and its appointed or elected leaders must be made to demonstrate loyalty to democratic ideals and friendship with the United States. By allowing a hasty creation of a new indigenous government, we risk empowering factions and institutionalizing corruption that will undermine U.S. military control and lead to political failure in achieving our war aims.

The ill-considered, rapid creation of new governments unworthy of local, international, or American popular support has been the major failing of American foreign wars since World War II. For instance, Ngo Dinh Diem in South Vietnam and al-Maliki in Iraq led their respective countries into ever-deeper sectarian and civil war, undermined the democratic reforms that we view as important to long-term success, and necessitated a broader and longer U.S. counterinsurgency effort that proved politically divisive in America. Unloading the national government responsibilities on the best available indigenous political faction simply may not provide the long-term solution we desire. Any attempt to impose/provide new governance on/for any nation can only be viewed as a long-term commitment of substantial U.S. governmental and military resources.

Most of America's experience and the mainstream professional research suggest that the bottom-up approach of restoring government first at local, then provincial, and finally national level succeeds best.[22] (The successful top-down approach used by MacArthur in Japan—the virtual co-opting of the intact Japanese government—should correctly be viewed as an exceptional achievement.) The danger in the bottom-up approach is that local leaders, once empowered, may seek independence or autonomy against U.S. desires. We should not assume that the population has a strong preference for reestablished nationhood and is willing to make local sacrifices to that end, as we found in Germany and Japan after World War II.

Most of our recent experience at nation-building in the former Yugoslavia, Somalia, Afghanistan, and Iraq has occurred in populations exhibiting mainly centrifugal, rather than centripetal, forces. Keeping these nations together while extending democratic institutions at all levels will tax the abilities of the most capable new indigenous government. Indeed, the nation's leaders may prefer renewed dictatorship that can maintain nationhood within the preexisting borders to allowing the breakup of the nation in response to democratically expressed regional desires. The twin challenges of preserving nationhood and democratization of the country's political system are not always compatible, so the military government division may simply be unable to find a formula for turning back political power to the occupied population that does not risk compromising America's war aims or democratic principles to some degree.

The United States should avoid, if at all possible, employing a simultaneous bottom-up and top-down nation-building approach, which characterized our wars in Vietnam and, so far, Iraq. Both of these efforts began by installing an indigenous national government that failed to live up to our expectations and could neither secure the nation nor extend the government's popular base. Rather than allow failure, Americans then bypassed the national government, going directly to local authorities in an attempt to improve support for the American effort. The result was a weakening of indigenous governance as the local authorities worked at cross purposes with the central government.

Often, the American perception that progress is being made is illusion. As one American authority in Vietnam commented:

> The officials of the host country are more often than not harassed, underpaid, and bewildered in the face of new problems. If they cannot avoid frequent confrontations with eager and demanding American counterparts, they tend to resort to supine acquiescence (which is rarely translated into action), stonewalling, dissembling, or playing one American official against another. We have learned, or should have learned in Vietnam, the futility of trying to cajole officials into pressing forward with American-sponsored programs that are not actively supported by their own government.[23]

After a point, American meddling and micromanagement create a three-way fight between the military governors, the national government, and local authorities that often serves only to undermine the legitimacy of all three.

Building Indigenous Security Forces

The new military governance division would have sections dedicated to the rebuilding of the nation's security forces: its defense ministry, armed forces, and national police. Profound policy choices would have to be made, and many of the policy options would have dramatic impacts on army occupation operations, manning, and budgeting.

Perhaps the simplest and most alluring decision would be to retain the existing armed forces and national police in place and simply purge their leadership, but rarely would our opponent provide us with standing forces that could be so easily co-opted. Historically, few occupiers have elected this route.[24] Generally, the defeated security forces are immediately demobilized and then re-created over time by the occupier in a different form under new leadership. This can only be seen as a race against time. The occupier's cost of providing security, especially maintaining law and order, is high and his efficiency low. Even though the occupier must achieve an initial monopoly of violence, he cannot sustain it for a prolonged period without increasing popular resentment, probably leading to insurgency or other forms of resistance. There is a need for both short-term programs, designed to provide immediate manpower and capabilities to the occupier, and longer-term programs, designed to produce the "objective" national force. Both programs would necessitate considerable prewar preparation and a partial reorganization of the army to accomplish.

Historically, immediate help has been procured in many ways. Local police forces have been partially disarmed but retained under military police supervision to enforce local law and order (as in Germany and Japan). Army tactical units have been reorganized and retrained as constabulary units to assist in population-control duties while provincial and national police units are restructured and purged (as in Germany). Local militias have been formed with integrated and embedded U.S. service personnel to perform local security (as in the Combined Action Platoons in Vietnam). Native scout units have been formed under U.S. command to assist U.S. tactical units in surveillance and reconnaissance (as in the Indian Wars and the Philippines). United States platoons have been augmented with local soldiers replacing some U.S. soldiers (as when the Republic of Korea soldiers served in U.S. Army units in South Korea; Korean augmentation to the U.S. Army was known as KATUSAs).

Some or all of these approaches should be considered and planned before the occupation even begins. Many of these approaches require that nonstandard units be created and trained even prior to the commencement of the major combat actions that enable the occupation.

Longer-term reconstruction of the military is completely dependent on local conditions. Germany formed the Bundeswehr in 1955 using German military expertise and available equipment, mostly American. The Germans proved very resourceful in their own rearmament, developing their own industries and doctrines, with the understanding that the new German armed forces would be used solely for the defense of Germany and commanded by NATO.

At the other end of the spectrum are Iraq and South Vietnam, where the U.S. Army created new national armies along American lines, using American-provided equipment and trained by American advisors. These advisor teams represent a significant claim on the army's force structure. In Iraq in May 2007, for instance, over six thousand advisors organized into over five hundred different military and police transition teams were assisting in the training and employment of the Iraqi Security Forces.[25] In addition, an entire brigade combat team at Fort Riley, Kansas, has stood down from its combat mission and been retasked to train army advisor teams destined for Iraq and Afghanistan.

All in all, the advisor effort represents about ten thousand total active-duty army personnel spaces, not one of which was part of the prewar force structure. With no doctrine or plan on hand in 2001 or 2002 to produce the large advisory structure that would be required, the army has struggled to this day with ever-changing makeshift arrangements for providing advisors, predictably harming the effort.

Making matters worse, the army personnel promotion system does not recognize the advisor assignment as "key and developmental," meaning that many of the most talented and ambitious officers will try to avoid these assignments. A recent Command and General Staff College (CGSC) graduate e-mailed from his new advisor assignment in Iraq:

> max boot was right. army has assembled the army of dud rounds for mitt teams. i've seen a team mutiny four times and be prohibited for a time from conducting its advising mission due to complete and utter dysfunction. i've got my own problems on my team, with captains so junior and immature as to place the team squarely at the margins of dysfunction. and there's hardly a team i've seen that does not have very significant leadership problems, at battalion, brigade, and division team levels. and there's no replacements for the dud rounds under the current model of filling mitt teams. we're all stuck with our 11 selves for 12 months bog plus tdy at riley prior to that, mutiny will not change that. if i were to characterize the leadership challenges, i'd say that not one member of the team can handle the requirement to be part small unit leadership, and a significant major part of the job is akin to organizational leadership.

Another recent CGSC graduate e-mailed from Iraq in August 2007:

> Also, beat them down on the FID [Foreign Internal Defense], we have enough to handle with ill-trained MiTTs, BiTTs, and SPiTTs [military, border, and special police transition teams]. Engagement is the name of the game so prepare to spoon . . . with your Host Nation/coalition partner (Iraqis).

One of America's top counterinsurgency experts recently proposed a twenty-thousand-man Permanent Army Advisor Corps, even if it meant reducing the number of army brigade combat teams focused on force-on-force combat, so critical is the advisor mission in most cases.[26]

Modeling the Rule of Law
"Revolutionary soldiers must be tried and punished where they committed their crimes."[27] American soldiers participating in an offensive war or an occupation are

revolutionary soldiers, part of an effort to reshape a national identity and build a new system of governance based on the American model. Success in this mission requires the highest standards of conduct and discipline. Revolutionary commanders from George Washington to Mao Tse-Tung have recognized that the people will judge the righteousness of their cause largely by the behavior of their soldiers. Consequently, ill-disciplined soldiers—and there are always some—need to be brought to justice in the same communities in which they committed their crimes so that their victims can see justice done.

Our record in Iraq has not been good. All too often soldier offenses have not been tried at all. When they have, the courts-martial typically occur years after the fact on an American military installation in the United States or Germany. Under these circumstances few Iraqi witnesses make the trip to testify. Fewer still will come to witness the trial. The juries are composed of U.S. military personnel who are understandably loathe to convict, often because of the incomplete evidence presented in such a trial, but more often because their remoteness, both in time and space, from the alleged crime makes it difficult to determine beyond a reasonable doubt whether the accused acted reasonably or criminally given the circumstances and the danger. Typically, these trials end in acquittal unless there is a confession or American eyewitness testimony.

American contractors, such as Blackwater, are not subject to military law, the Uniform Code of Military Justice, and so have not been held legally accountable for seemingly criminal conduct against Iraqis. Early in the occupation, Coalition Provisional Authority Paul Bremer exempted them from punishment under Iraqi law. Few contractors, if any, have been tried in American courts for alleged offenses against Iraqis or Afghanis.

The American practices concerning the disciplining of its armed forces through the rule of law have proven inadequate in Iraq, one reason why the Iraqis insisted in shared legal jurisdiction over U.S. servicemen and complete jurisdiction over American civilians in the December 2008 status of forces agreement (SOFA). We can do better.

As a matter of policy, American servicemen who are part of an invading or occupying force and are suspected of crimes against persons or property in the invaded and occupied county should be tried and, if convicted, punished, at least for a while, in that country. Service members must be told beforehand that their quickest way home is honorable and legal service; criminality will keep them overseas. The same policy should apply to contractors and other civilians accompanying the force. All should be subject to applicable articles of the Uniform Code of Military Justice as directed by the commanding general of the invasion or occupation force and suffer the same punishments if convicted by court-martial. Juries should be constituted from soldiers serving in the same area under the same rules of engagement to better determine whether the action at issue was reasonable or not, given the situation and the mission. Trials should be public, insofar as possible, both to model the American system of justice and to assure the population that the trials are fair.

Such a policy would not be a departure from American military tradition. General Washington used such a policy during the American Revolution. Winfield Scott used it in Mexico in the 1840s. The Union Army tried its lawbreakers in front of Southern witnesses. The American armies occupying the Rhineland after World War I and Germany and Japan after World War II regularly tried, convicted, and punished soldiers in those countries for offenses against German and Japanese citizens. Rule of law is what elevates the U.S. Army above the Wehrmacht, the Red Army, or a host of other militaries that try to establish their rule by terrorizing foreign populations.

The rule of law works both ways. We cannot expect the invaded country to model itself after us if we treat its citizens cruelly or criminally. America must field a disciplined force, and military policy must demand that the few criminal actors who find their way into our armed forces be tried and punished with due process that satisfies both American legal requirements and the population's justifiable desire for justice. American servicemen made aware of such a policy and trained accordingly would better serve their country and their consciences.

The central idea in this chapter is simply that prior planning and organization can improve military performance—indeed, often make the difference between victory and defeat. By standing up a military governance division prior to the force-on-force campaign, many of the postwar policy decisions can be considered and decided before the war ever begins, and the necessary force structure for postwar occupation and reconstruction can be in place when it becomes needed. Rather than assuming, as we did in World War II, that the war will last years and the critical decisions can be deferred until after the war begins, we should instead assume that the war will be short and that critical postwar decisions and preparations must be made in conjunction with the prewar planning.

The situations in Afghanistan and Iraq could be far different today if we had had two military governance divisions in the force structure on 9/11. One could have stood up for Afghanistan. The other, beginning in December 2001 when CENTCOM began its OIF planning, could have focused on Iraq. The right people for postwar governance could have been on hand for a year prior to the Iraqi invasion, important policy choices could have been made with full consultation, force structure could have been created, units could have been trained, and execution could have been war-gamed and rehearsed. Instead, none of this happened and we ad-libbed our way into a "fiasco" from which we may not recover.

I do not wish to claim that a correctly planned and executed occupation would have prevented all insurgent and sectarian violence in Iraq; after all, by their nature, wars involving governance are wars of attrition. Perhaps, however, we could have reduced the dimensions and duration of the violence. We owed it to ourselves and the Iraqis to have at least tried to limit the carnage. We also owed it to ourselves and the Iraqis to have more carefully thought through the mechanisms of popular governance that would have satisfied our mutual interests, rather than allowing events to fracture the Iraqis into warring camps. Having failed to do so in Iraq, it is not too early to prepare, if only intellectually and doctrinally, for our next offensive war.

Chapter 11

Declaring Offensive War: Implications for Civil-Military Coordination

Offensive war is almost always elective; rarely is it truly forced upon us as being the only policy option available. For instance, Japanese and German actions in December 1941 forced us immediately into a defensive war, but the "unconditional surrender" policy—Roosevelt's declaration of an offensive war of occupation—did not come until early 1943. We and our Allies could have forced a conclusion to World War II without occupying Germany and Japan; however, we chose not to. Indeed, the Allies constantly conferred with one another to preclude one party or another agreeing to a separate peace.

Similarly, we could have fought the global war on terror without invading Iraq; however, once again we chose not to. Even if attacked, we can always try to limit a war to not involving questions of governance of the attacker, as we did with the Chinese and ultimately the North Koreans during the Korean War, and with Iraq in the Gulf War. It is not automatically true that all wars must end with the demise of one regime or the other. During the Vietnam War we never sought to change the North Vietnamese regime. Rather, it is more generally true that the initial attacker's decision on whether to seek regime change is made before war begins, while the initial defender's decision about whether to seek to destroy the initial attacker's regime is generally made after the war begins. In other words, either side may choose to try to arrange a limited war rather than seek the occupation and regovernance of the other.

Cost-Benefit Analysis

Whatever benefits may be envisioned from an offensive war of occupation must be weighed against the costs of that military choice. History provides precious little

the "entry phase" of the war. Simply put, the forty-eight brigade combat teams and their supporting organizations that already exist in the army force structure, not to mention the additional marine expeditionary brigades, are an overmatch for any conventional force in the world. We need not build more of those units for offensive war; we need to create units designed for governance, occupation, and reconstruction so that we as a nation can take advantage of the fleeting moment that the combat units offer to impose "regime change." Given a year's foresight of the intent to launch an offensive war, similar to what we enjoyed in Operation Iraqi Freedom, one should expect that America should be able to mobilize and train the appropriate forces.

Getting Indigenous Governance Right

American policymakers must pay much greater attention, before the war even begins, to the regimes we propose to install in foreign countries. As Field Manual 3-24, "Counterinsurgency," repeatedly states, there is a clear link between poor governance and the development of insurgency. As the occupier in an offensive war, we have a tremendous interest in minimizing, insofar as possible, insurgency against our military governance and the local governments we empower. Unfortunately, we have too often invaded first and then ad-libbed our military governance or, worse, hastily installed indigenous regimes unworthy of either local or American domestic support, and then fought to keep them in power. Our abrupt and ill-considered elevation of Ngo Dinh Diem to the presidency in South Vietnam in 1954, after an equally hasty and flawed decision to support the restoration of the French colonial regime in 1945, led to decades of costly war that perhaps could have been avoided had we devised governmental arrangements that the people of Vietnam could have welcomed. The al-Malaki government in Iraq, like that of Diem in Vietnam or Reza Pahlavi in Iran, also suffers from limited support in Iraq and, on and off, among American opinion makers. A creation of poorly considered American occupation decisions, the al-Maliki regime nevertheless has become the crucial determiner of whether the American effort in Iraq is successful. How much of the bloodshed in Iraq is a reflection of al-Malaki's policies is unknowable, but perhaps a better alternative could have been found.

It is notable that in our most successful occupations, post-World War II Germany and Japan, the decisions about what sort of new indigenous governance would be created were the result of years of deliberation and research at the highest level of the U.S. government, as well as a general desire to genuinely liberalize and democratize the conquered nations. These well-considered policies, as well as competent military governance, gained us the goodwill of the citizenry of the two occupied nations so that insurgency never developed. Our more cavalier, less responsible governance solutions in Korea, Vietnam, Iran, and Iraq all bred insurgency toward the government and hostility toward Americans.

We must come to appreciate that, at least for the United Sates, governance decisions are the most important policy decisions in warfare and will come to define all the military actions that come before or after them. Consequently, every effort has to be made to get the governance right. Expedient options must be seen for the traps they can become. Only the long-term interests of the American citizenry and the citizens of the nations immediately affected should be considered. Once the war concludes, Americans have a vested interest in seeing that the occupied populations are kindly disposed toward their new governance. Only deliberate policymaking and military actions will make that possible.

The Army Must Write Occupation Guidelines

Based on the historical record, the army cannot rely on the incumbent administration to give it timely guidance on occupation policy. It must either force a presidential decision or, more likely, produce a policy on its own professional initiative and institute it as a fait accompli.

The hue and cry that Scott's General Order 20 caused is illustrative of the Washington process. Scott drafted his legal code for the Mexican War prior to even being appointed its expeditionary commander, and then forwarded it to the secretary of war for approval in late 1846. Secretary Marsh refused to act on the matter and returned it to Scott as "too explosive for safe handling." The matter went to Congress, which debated, on philosophical grounds, whether any legal code should restrict the actions of the American army in Mexico. Some argued that the "laws of nations" conferred upon the stronger nation unlimited prerogative. But the majority agreed with Congressman Seddon, from Virginia, who eloquently summarized a more enlightened position for America:

> It is the boast of modern times . . . that in war, as in peace, a code of law to govern all international relations, founded in part on the practices of nations, but more correctly binding, as deduced from the most sacred principles of justice and the highest ethics of morality and humanity, has, by general comity and common convictions, been established and recognized. . . . The worst of all conditions for a people is to be without government at all—a prey to anarchy and confusion, with their rights, their property, and their persons, at the mercy of the ruffian or the ravisher, whose excesses no law restrains and no justice punishes. For a conqueror to overthrow an existing polity, and to leave a submissive people to such horrors would be such a tyranny as no principle of humanity and law could tolerate.[1]

Still, the Congress could give Scott no practical guidance, only the advice that American commanders in Mexico should be guided by fundamental moral law. As far as writing occupation policy and law was concerned, Scott was on his own.

Washington's performance has not improved since 1846. Not until one year into the Civil War, in 1862, did Lincoln appoint a commission to write a code of conduct for Union soldiers in the occupied Confederacy, and by the time General Orders 100 was approved in April 1863, the war was half over and much of the South was already in Union hands. Until that time, the army and its commanders in the field had to use their own judgment concerning occupation policy.

Washington provided virtually no guidance to the army for the occupations of Cuba and the Philippines that resulted from the Spanish-American War. Having declared the war and authorized the troops, Washington assumed that the army would know what to do, if it even gave the matter that much thought. Again army commanders on the ground, often guided by Scott's Mexican War example and the Civil War's General Orders 100, had to determine the policies and take the political heat of the inevitable second-guessing that went on in the press and the political arenas.

Even in World War II, America's paradigm achievement in occupation organization and policy, Roosevelt was slow to approve the army's initiatives that helped create the new world he desired. Roosevelt initially opposed placing the army in charge of postwar governance, favoring giving the task to the ill-equipped State Department, and was angry when he learned that the army had established the School of Military Government at the University of Virginia.[2] In September 1944, when shown a copy of the army's *Handbook for Military Government*, which had been prepared for military governance units in Germany, Roosevelt called a press conference to publicly denounce it. After minor modifications, the army distributed it anyway.[3] A less responsible army leadership might have accommodated the president and "gone with the flow," but their professionalism would not allow it.

In our post–World War II offensive wars—Vietnam, Iraq, and Afghanistan—the U.S. Army and Department of Defense relegated "ownership" of occupation and government-building policies to other U.S. cabinet departments, partly due to administration desires, partly due to professional myopia and wishful thinking. Chaos was the result.

Occupation policy and procedures, including the lead for nation-building, must be considered an integral element of the U.S. Army's professional expertise. Only the army can do it; no one else understands it; and no political body, either Congress or the administration, will provide timely and adequate guidance to the military on the task. The army must instruct the American government on occupation matters, not the other way around.

The Need for Political Will

For nations initiating offensive war, the need for political will to see the war through is paramount, especially once the occupation begins. Offensive war is not a sideshow, something that deserves secondary effort in the national life. Rather, it is an enterprise that wagers the national well-being, not just during the current administration but perhaps for generations to come. No offensive war should be taken lightly, sold deceptively, or waged divisively. National unity is imperative, but to merely plead for unity is simply idle. There must be mechanisms put in place that would to the greatest extent possible ensure supermajority agreement that the proposed offensive war is in the abiding national interest and that when the bills come due, both in lives and money, they will be paid.

First of all, we as a people should insist that any offensive war be accompanied by a declaration of war in accordance with the Constitution. Second, the mobilization of the military governance divisions and other units necessary to prosecute the offensive war to victory should be allowed only by act of Congress, perhaps by a supermajority of Congress. The national representatives most answerable to the people must be forced to declare their position on the war and be made responsible for raising the army to fight it. Placing too many prerogatives in the executive branch may get the war started without ensuring the unity needed to see it through to the finish.

We should never begin an offensive war unless we are willing to build the force structure and political foundation necessary to prosecute the war to final victory: the empowering of a new sovereign government in the defeated country that will be an ally for the foreseeable future. This will most likely require a decade-long commitment to the war, the occupation, the resulting counterinsurgency, and finally the building of stable governance. Failure will likely result in anarchy followed by the emergence of a new government hostile to our interests. It could lead to even worse consequences—the formation of an alliance against us or to lasting bitterness that will bring war to succeeding generations of Americans. I am not an advocate for offensive war; I simply believe that, should it become necessary, we enter the endeavor with our eyes open and our preparations well made. There is no form of war more unforgiving of error and more punishing of success than offensive war.

Calls for Regime Change: Offensive War on the Cheap

There has developed an unfortunate tendency in American foreign policy to label certain hostile governments as anathema, cursed by their very nature and in need of fundamental change. Communist governments, especially the Soviet, Communist Chinese, and Cuban, were viewed this way since their formation.

After the Cold War, calls for regime change shifted to Iraq. President Bush's Axis of Evil formulation further put Iran and North Korea on notice. American spokespeople often proclaimed that all nondemocratic governments worldwide must change their ways, that they are flawed and on their way out. I am a firm believer in liberal democracy and the rights of man; however, I fail to see why we would demand a rather immediate democratization of societies where liberal appreciations have not yet taken root. It is a matter of time frame. If it took the world three hundred years to get from the Enlightenment to its current level of democratization, why is it unreasonable to expect that the further democratization of the planet will also be a matter of centuries? If the spread of democracy so far has been largely the free choice of the societies so governed, why do we think that now is the time to compel societies as yet unconvinced? To use Francis Fukuyama's famous teleological term, why must we force the end of history?

From a military standpoint, rhetorical calls for regime change are offensive wars on the cheap, eliciting popular revolt to overthrow the local government without a U.S. military occupation. One should expect the identified regimes to become defensive and take all measures necessary, internally and internationally, to ensure their survival. In this respect, the U.S. government's overt branding of a government as evil may be counterproductive, casting its internal foes as agents of an external power and legitimizing state repression. The government may also elect to strengthen its external defenses and alliances.

All of America's loose talk about invading Iran in 2003 has only hardened the anti-U.S. positions of the Iranian government, led to an acceleration of its nuclear weapons program, and moved the country closer to China and Russia diplomatically. In a counterproductive way, calls for regime change through popular revolt make regime change less likely.

In their combined religious imagery and anticipated U.S. commitment, recent calls for regime change echo a Manichean belief that evil destroyed is goodness created. Were it only so simple, Earth would have long ago become paradise. Perhaps we should more fully appreciate that internal revolution can not only provide an opportunity for better government, but also for worse. One could hardly argue that the Russians were better off under the Soviets than they were under the tsar, or the Chinese better off under Mao than they were under the Nationalists, or the Iranians more free under the mullahs than the shah. Is Russia a better international partner under Vladimir Putin than it was under Mikhail Gorbachev? Will a post-monarchy Saudi Arabia become a liberal haven in the Middle East or a Wahhabist terrorist state? All too often, bad replaces bad when a society has not embraced liberal ideas. Regime change through popular revolution provides only an opportunity for change but no certainty that the change will be positive.

Iran illustrates the Pandora's box perils of American-sponsored regime change. Allied occupation of Iran in World War II led to American support for Shah Mohammad Reza Pahlavi as a constitutional monarch. When Iran's popular postwar prime minister, Dr. Mohammed Mossadegh, nationalized the nation's British oil concessions, the American government conspired with the shah for Mossadegh's ouster by the military in 1953. The shah, with strong American support, reversed most of the oil nationalization and ruled Iran as a secular police state monarchy for the next quarter century. Never popular, Pahlavi's eventual ouster by the Iranian revolution of 1979, an improbable coalition of religious fundamentalists and liberal reformers, ushered in an era of unprecedented anti-Americanism in Iran. Whereas Americans tend to see Iran's challenge in terms of Iranian policies toward religion, Israel, terrorism, nuclear weapons, and the free flow of oil to world markets, Iranians view their behavior, in part, as a nationalistic reaction to the American imposition of an unwanted ruler on the Iranian people.[4]

The long shadow of the Mossadegh coup continues to poison relations between Iran and America and, from the time of the Iranian revolution in 1979 to this writing, has led to hostile relations and episodic violence between Iran and the United States. Many Americans in and out of government have recently advocated renewed American military attacks against Iran and yet another American-sponsored regime change in that nation, despite the lessons of history. The ouster of Mossadegh, however attractive it may have seemed at the time, was a tragic U.S. error from which we and Iran have not yet recovered.

Neither has the United States demonstrated an ability to guide revolution and regime change, once it has spontaneously occurred, without significant U.S. military involvement.

The most egregious example of this was our failing to guide the democratization of Russia following the collapse of the Soviet Union. Here was an event that we had predicted and patiently worked to achieve for over forty years, yet when the time came we could not assist Russia with effective policies and support that would bridge the chasm between its totalitarian past and a liberal democratic future. Instead, our ideologically driven insistence on "shock therapy" led to widespread looting and criminality in Russia, a loss of popular faith in the West as a model and friend, and the current Putin retrenchment into old Soviet ways.[5] All of the advisory efforts of the U.S. government, allied governments, and the private sector could not prevail in Russia, a profound failure the true cost of which we do not yet know.

An American military governance division might seem the ideal organization to steer a new revolutionary regime in positive directions, but such a thought is illusory. No revolutionary government, whether seizing power through political or military force, is likely to invite Americans in to co-opt their rule, no matter how highly they regard the American system. Should the American military appear uninvited, we will be opposed. The reaction of Aguinaldo and the Filipinos to America's offers of "aid and good government"—that is, armed defense of their sovereignty—is only one example of our unappreciated interventions in foreign revolutions. Our misguided expeditions into Russia in support of the Whites after the 1917 Communist revolution were ineffective and, in the words of Gen. Peyton C. March, the army chief of staff from 1918 to 1921, "a military crime."[6]

Neither could we prevent renewed civil war, and ultimately the Communist takeover, in post–World War II China, despite the best efforts of Gen. George C. Marshall and a corps of American marines.[7]

Indeed, there is no demonstrated American civilian or military mechanism for guiding regime change once it occurs spontaneously in a foreign country. Unless we are willing to invade and occupy, meaning paying the cost of an attrition war, we must accept the fact that the processes of revolutionary regime change and the development of new governance are largely outside our control. Given these circumstances, American calls for regime change are generally irresponsible, often counterproductive, and always fraught with peril.

Placing Due Emphasis on Governance

Unless our goal is to exterminate our enemy in a campaign of mass murder, imitating Hitler, or drive him out of his homeland, imitating Milosevic, we must eventually govern him. And unless we intend to forever pay the high costs of governing a subjected population against its will, we must someday allow that population its own sovereign government. Yet the U.S. military currently has no specific doctrine or organization devoted to military governance of occupied areas or post-occupation transition to home rule. The prevailing doctrine and force structure are designed to support host-nation governance, not to provide governance to conquered people. This state of affairs is the single greatest weakness in our military today.

I have already suggested the creation of a military governance division that would be mobilized for offensive wars to provide governance in occupied countries. Attempting to govern through tactical commanders, even if they are augmented by civil affairs staff and civil affairs teams, simply doesn't work. The current army system in Iraq is an ever-changing, crazy-quilt patchwork of doctrine and force structure that is ill suited for offensive war.

The U.S. Army as an institution must elevate military governance commensurate with our renewed interest in offensive war. The doctrine, Field Manual 27-5, was last published in 1947 and has been officially obsolete for decades. For a few years in the 1950s, the army's Provost Marshal General School's military government department kept the doctrine alive.[8] Field Manual 41-10, "Civil Affairs Military Government Operations," published in 1957, was the last doctrinal publication to cover the subject. By the time the manual was updated in 1962, named only "Civil Affairs Operations," military governance units had disappeared from the army (in 1959); their personnel had been reassigned into civil affairs units. Thus began the "ultimate demise of [U.S. Army] overseas constabulary doctrine."[9] In 1964 the army's Civil Affairs Agency was merged into the Special Warfare Group.[10] In 1971 the Civil Affairs School at Charlottesville, Virginia, was closed and moved to Fort Bragg, North Carolina, where it was absorbed by the John F. Kennedy Special Warfare Center and School.[11]

The smaller post-Vietnam army, all volunteer and focused on the defense of West Germany, envisioned little need for civil affairs units and in 1977 announced decisions to dramatically reduce the civil affairs force structure, both active and reserve. By the 1980s, civil affairs had virtually disappeared from the army's capstone doctrinal manuals and training exercises.[12] In 1990 civil affairs was lumped together with psychological operations with the creation of the U.S. Army Civil Affairs and Psychological Operations Command (USACAPOC). In 2006, USACAPOC, it turn, was shuffled off to the U.S. Army Reserve Command.[13]

The sad descent of the army military government capability from its apex in the post-World War II occupations to its current oblivion—"military government" is not even a term we now recognize in our doctrine—correlates to our growing inability to favorably conclude offensive wars with acceptable political outcomes. We must urgently reverse both trends. The crucial doctrine must be reviewed, updated, and republished; the necessary force structure must be created. Some organization needs to be made clearly responsible for creating America's new military governance capability. To affix personal accountability, military governance needs to be somebody's primary mission, not everyone else's afterthought, if we are to successfully conclude offensive wars.

I think it is necessary that military governance, and civil affairs in general, must stand up its own army school and center and be removed from the special forces community, where civil affairs now resides as a bright red-headed stepchild. The ethos of U.S. Army Special Forces—the ideal of a small, highly trained Green Beret A-team in the middle of nowhere working almost autonomously—is simply incompatible with the mindset needed to provide governance. Military governance is not built on small, independent teams but on large bureaucracies and supporting organizations, such as police forces, courts, and chambers of commerce. Military governance requires large concentrations of civil affairs officers working at the seat of government, smaller concentrations in provincial capitals, and dispersed teams at local levels, but all still strictly subordinated to central policy and control. Neither is governance a clandestine or "special" mission. Indeed, from the civilian's perspective, military governance is the most public and normal mission the army ever does. Governance is not an A-team mission. Nor is it a mission accomplishable by a scattering of small teams working in support of local tactical commanders.

Nor does military governance, or even civil affairs, lend itself to the standard personnel policies employed by special forces or the army in general. Our cloistered military, isolated on rural stateside posts and intensely trained within its own community, will never be able to grow enough military governance expertise within its own ranks to properly govern an occupied country. Getting enough of the right civilians into uniform to accomplish the task will be the primary mission of my proposed military governance division when the time arises for it to mobilize. Direct commissioning of individuals with the necessary expertise is the only practical way for the army to procure the right people.

It makes sense to me to reestablish a center and school for military governance and civil affairs at Charlottesville, Virginia, and align it with the army's Judge Advocate General School, as it was during World War II. Governance and civil affairs have much more in common with the legal profession than they do with special operations. It is not by accident that many of our most successful military governors, Winfield Scott in Mexico and Franklin Bell in the Philippines among them, had the benefit of legal training. The liberal environment of the University of Virginia, contrasted to the strict military character of Fort Bragg, would be more conducive to assembling the right people with the talents necessary for the governance mission. Proximity to Washington, D.C., and the various agencies of the federal government is an added advantage of the Charlottesville location.

Chapter 12

The Frontiers of
Strategic Offense

As American ground forces—soldiers and marines— struggle to occupy Iraq and Afghanistan and produce lasting favorable governance, we should well consider the limits of our current expansion. In the euphoria after the rapid capture of Baghdad, some neocon pundits argued that the U.S. military should just keep on going, turning either right into Iran or left into Syria.[1] These self-serving idealists, like William Randolph Hearst in 1898, seemed far more interested in building popular and political pressure to start wars than in thinking through the problems of actually waging or concluding the wars they urge. Fortunately the U.S. military, which had little idea of the scope and duration of the problems it had already signed up for in Afghanistan and Iraq, at least had the good instinct to realize that it had already bitten off as much as it could chew. Never was the immediate military exploitation of the success in Iraq into Syria or Iran seriously considered.

By 2007, however, some in Washington were again pushing for war with Iran, and many in America and the Middle East believed that war against Iran was both imminent and inevitable.[2] One would imagine that such a war might be limited in nature, aimed at halting Iranian interference in Iraq or halting its nuclear program, but given America's stated preference for regime change in Iran and its inclusion by President Bush in the "evil empire," one could not rule out that the same political pressure groups who favored our offensive war of occupation in Iraq might favor a similar American war in Iran. The new administration of Barak Obama is less anxious to engage Iran militarily, but seems more eager to take on the ungoverned spaces in Pakistan. All of this begs the question of how far out we can, as a practical matter, push the American military frontier. What *are* the limits of the current enterprise to bring governance to Thomas P. M. Barnett's "gap"? What are the costs and risks?

A Hypothetical Invasion of Iran

What would a 2007 invasion of Iran have looked like? Iran is a nation of some 65 million people about the size of Alaska, in both respects about three times the size of Iraq. Much of the terrain is mountainous, and, unlike Iraq, major mountain ranges separate the Persian Gulf from the major population centers in the county's interior. Though I have no personal knowledge of any classified U.S. military war plan regarding a war with Iran, its size alone would suggest that invading and occupying Iran would require at least six hundred thousand soldiers and marines, virtually all of the deployable force structure.

The logistics and engineering challenges involved in getting such a force from its staging area, presumably Iraq, to Teheran and other Iranian population centers would slow progress and provide opportunities for Iranian counterstrikes. The length and vulnerability of the supply lines would necessitate the commitment of substantial follow-on combat forces to secure rear areas as well as service and support units. American ground forces would eventually make it into the densely populated regions around Teheran and Isfahan. Following the pattern of the recent Iraq War, the American invasion force would be able to successfully fight its way to the interior and then be confronted by irregular forces fighting a guerilla war.

Unlike Iraq, most of the anti-American energy will not expend itself in sectarian conflict between Shiites and Sunnis but will find its outlet by attacking and inflicting casualties on U.S. forces. Based on precedents from Vietnam and Iraq, we could expect the counterinsurgency effort to consume at least half a million U.S. troops for perhaps five years.

Based on the bottom-up example of the American zone in post–World War II Germany—12,000 military governance troops for 25 million Germans—establishing new governance in Iran would require perhaps 25,000 military governance troops especially trained for occupation duty. (Using the World War II planning factor of two constabulary troops per thousand people, a 130,000-man constabulary force would have to be created to provide police services for the occupied population. The entire package would total about 750,000 men.)

The overwhelming concentration of U.S. military forces in Iran would lead to vulnerabilities elsewhere. At a minimum, the U.S. military's reserve forces would have to be fully mobilized to mitigate the various global risks created by our invasion of Iran.

The cost of this effort would be staggering. American deaths would probably reach into the low tens of thousands, with perhaps a hundred thousand or more seriously wounded or injured. Iranian deaths, predicted at 5 percent, would come to over three million. Perhaps enlightened counterinsurgency policies could reduce that figure to a million or so. Millions more would be homeless. The financial cost of the war would reach several trillions of dollars, perhaps more, based on Operation Iraqi Freedom costs.

Worse, any American attack would surely be answered by Iranian attempts to sabotage or otherwise destroy oil production and export capabilities throughout the Persian Gulf, some of which might be successful. Oil production has a vast and fragile infrastructure; fighting among it will almost certainly decrease production.

Tremendous geopolitical risk would accompany the war. China and India, both of which have signed large oil and gas deals with Iran, would not be pleased to see Iran's one hundred billion barrels of oil reserves fall under American domination. How they would respond is unknowable, but neither country has an interest in seeing an American war in Iran succeed.

Russia, too, would be tremendously upset by the American military advance to the Caspian Sea and our consolidation of a contiguous frontier in central Asia. Like China and India, Russia's response is not now predictable, but its options—beside direct military action against American forces—include arming Iran, seizing parts of central Asia, perhaps blocking or seizing portions of the Baku-Tiblisi-Ceyhan pipeline, or cutting off oil and gas shipments to Europe and the West as a way to increase pressure on U.S. policy.

In the longer term, there is the distinct possibility that other nations of the world could combine in alliance against the United States. As America is perceived as becoming bolder in its disregard of others' national sovereignty—first Iraq, then Iran—others may take more aggressive measures for their own defense.

Little of this has been discussed by proponents of war with Iran. They focus on the Iranian threats to Israel, its support of Shiite militants in Iraq, or its support for terrorism in general. Rarely do the war hawks try to quantify the actual harm that Iran does to the United States or discuss alternative means of handling these problems. War is the preferred solution. Some supporters of such a war predict, as they did prior to the Iraq War, that American forces would be greeted as liberators in Iran, where people are thought to be desperate for freedom. There is no data to back this extraordinary belief. The reality is that an invasion of Iran would commit the United States, on a purely voluntary nature, to a long and difficult war, followed by the largest nation-building endeavor in our history. Whether the identifiable gains are greater than the identifiable costs would rationally be determined by careful analysis based on the coolest of judgments, not the heat of inflamed rhetoric. Factored in, too, must be the risks that we would assume to our long-term interests, and to those of our allies and friends, by initiating war.

Defense Through Nuclear Weapons

Are there candidates for invasion and occupation other than Iran? What would such countries look like?

First of all, and this is a point not lost on the rest of the world, these countries would lack nuclear weapons. No country possessing nuclear weapons has ever been invaded by another country. This has proven true since 1945 and seems unlikely to change in the foreseeable future. Possession of nuclear weapons and delivery devices, even in small numbers, makes any potential enemy reluctant to attack. No offensive war imagined to date would have yielded sufficient gains to compensate the attacker for the loss of even a handful of his key metropolitan areas to nuclear bombings. Those initiating aggressive war must convince their people and themselves that the homeland will never be endangered by the war; nuclear weapons and long-range missiles or bombers make this assurance impossible.

Shorter-range theater nuclear weapons also serve to deter aggressive war. The logistics bases and troop concentrations required for the assembly of an invasion force are completely vulnerable to nuclear-tipped missiles and other weapons of mass destruction. Had the Germans possessed tactical nuclear weapons during the Normandy invasion in June 1944, the Allied armies would have died on the beaches and their naval support been sunk at sea. The fiasco would certainly have ended the ambitions of the Western allies to liberate France and occupy Germany. Similarly, neither Desert Storm nor OIF would have been attempted had Saddam Hussein possessed a credible nuclear force to use against U.S. forces massing in the Gulf.

Indeed, to pursue an offensive war strategy, the United States must ensure that none of our likely targets achieves nuclear capability. To assure themselves of adequate deterrence against American military attack, the small nations opposing perceived American hegemony know they need nuclear weapons. In this respect, the 1981 Israeli bombing of the Osirak nuclear plant in Iraq, our imposed inspection regime in Iraq after Desert Storm, our invasion of Iraq in OIF to dismantle their allegedly hidden nuclear program, our previous threats against Iran, and the 2007 Israeli bombing of an alleged nuclear plant under construction in Syria can all be seen as part of the same offensive strategy.

We claim the right to keep these countries nonnuclear, at least in part, so that we can preserve the right to change their regimes, by force if necessary. They try to develop a nuclear capability, at least in part, to protect themselves from American attack and occupation.

The situation on the ground is more complex than the theory. Theoretically, American precision munitions can destroy small nuclear forces, but not with 100 percent certainty. Nations such as Iraq prior to OIF may bluff a capability they really don't have, hoping that doubt alone will prove sufficient to give America pause. Nations intent on developing nuclear capability camouflage their efforts and emplace key facilities in deep earth caves to shield them from American conventional precision strikes. Whatever the calculus, bona fide nuclear nations are shielded from an American offensive war strategy.

Too Big to Occupy

Irrespective of their possessing nuclear weapons, other countries are simply too large or populous to occupy. Even a country as powerful as the United States could never occupy vast and populous countries such as China or India or Russia. Nor could we forcibly reoccupy our current allies in Europe and northeast Asia. There are simply limits to military power that even a superpower must observe.

Pakistan, with a population of 170 million, nuclear weapons, and strong links to a variety of radical Islamic groups, is sometimes considered to be an unreliable ally in the war on terror and, potentially, a powerful enemy should the Islamists take control. Still, no matter what Pakistan's future, American occupation of the country is simply out of the question because of its demographic mass.

Brazil (180 million) and Nigeria (150 million) are not considered military problems, but their mass alone would preclude occupation. Mexico (110 million), which President Taft actually considered occupying a hundred years ago and was twice invaded by U.S. military forces prior to World War I, is no longer a candidate for occupation, despite the fact that the drug cartels operating out of Mexico surely do more harm to Americans yearly than all the Islamist groups of the Middle East have ever done.[3]

The scope and cost of occupying the world's largest nations would consume the strength of even the most powerful military in the world. Japan, despite full mobilization during World War II, could not occupy China. Germany, similarly, could not occupy Russia. We have our limits, too. The largest and most populous countries of the world simply defy reasonable occupation, and their boundaries must be considered the limit of an American offensive war strategy.

Ungovernable Regions

There exists still another category of nations, often called failed states, which are so politically fractious that they simply cannot be governed in the modern sense. These are generally impoverished regions comprising tribes and tribal groupings with little experience with, knowledge of, or capacity for the various social covenants and bureaucratic methods necessary to administer a modern nation-state. Afghanistan is such a region, as is Somalia. Our inability to pluck Somalia from Barnett's "gap" and lead it into the "functioning core" illustrates the high cost and ultimate futility of trying to transform failed states into governable nations.

America's involvement in Somalia began in the last days of the George H. W. Bush administration in December 1992. Four years of civil war between various warlords and factions had finally resulted in the complete collapse of the Somali agricultural economy. At least 350,000 Somalis had died in the fighting and from starvation by the time the Western nations took notice. In late 1992 television news networks throughout Europe and America fed their viewers a steady stream of pictures and accounts of the unfolding human tragedy, building a global consensus that something must be done.

Pursuant to United Nations resolutions, American forces entered Somalia on December 9, 1992, to provide security and conduct humanitarian relief operations. Operation Restore Hope was an immediate success, and by spring 1993 the humanitarian crisis was over. Realizing that the famine was a direct consequence of Somalia's warlord politics, and not wishing to see the recent gains in nutrition reversed, in March 1993 the United Nations, with American backing of course, enlarged the mission from humanitarian relief to occupation, empowering a mostly American twenty-eight-thousand-man international force to disarm the warring factions, establish law and order, and build democratic institutions: nation-building writ large.

The UN project was immediately rejected by the most powerful warlord and clan leader, Mohamed Farrah Aidid, who quickly began actions to frustrate the United Nations efforts. By June 1993, Aidid and the United Nations were at war. When Aidid's forces killed twenty-four Pakistani soldiers under UN command, American forces, operating outside United Nations control, began hunting Aidid and his henchmen in the slums of Mogadishu. In August President Clinton deployed U.S. special operations forces to Somalia to get Aidid.

On October 3 and 4, in what became known as "Blackhawk Down" or the "Battle of Mogadishu," Aidid and his militias surrounded and relentlessly, if unsuccessfully, attacked a force of 160 U.S. Army Special Forces and Rangers, killing or wounding most of the force and shooting down two helicopters. A relief force of Americans, Pakistanis, and Malaysians ultimately rescued the surrounded force, suffering three dozen additional casualties in the process. Aidid's casualties numbered in the hundreds, if not thousands.[4]

Still, Clinton judged, there was little to be gained in Somalia that was worth fighting a war over. He quickly announced the end of the American hunt for Aidid and, some months later, his intent to withdraw the American forces. The UN mission consequently collapsed. Somalia remains in a state of constant civil war and intermittent famine; it has become a base for pirates and international terrorists, but nobody in the West cares to reinvade.

Clinton's judgment has stood the test of time. Trying to build the institutions of enlightened national government out of a diverse collection of warring tribes and clans is a fool's errand. Nothing could be gained in Somalia that would compensate the West for the military casualties and financial cost required to occupy it. Whatever interests America has in Somalia are best managed from afar, by surgical strikes or raids, perhaps by funneling money to one faction or another. There is no interest that justifies occupying and governing such a fractious land.

In Somalia and other failed states, we again see the limit of America's offensive military strategy: regions not just impossible to occupy, not just too costly to occupy, but simply worth too little to occupy.

Control Through Non-Military Means

Then there are the nations that oppose American interests that are easily controlled in nonmilitary ways. Cuba, our nemesis in the Caribbean, or Venezuela, the home of the bellicose Hugo Chavez and the source of over a million barrels a day of American oil imports, are two such nations. Though at times both of these countries claim to be threatened by American invasion, the reality is that our diplomatic and economic levers of power are sufficient to keep their behavior within acceptable limits. In the aftermath of the Cold War, Cuba has become just another unimportant Caribbean island. Although Americans agree that a progression from Castro-ism to freedom would benefit the Cuban people and America, there is simply no reason to use military power to effect that transition. Americans are confident that, given the fullness of time, events will steer Cuba in beneficial directions. A recent Obama administration decision to open up travel to Cuba is perhaps a first step in the normalization process.

Similarly, despite his constant rhetoric regarding an imminent American invasion and weapons purchases, Chavez has little option but to continue shipping oil to America, there being no infrastructure developed to ship it anywhere else. The fragile Venezuelan economy would collapse without the revenues generated by the oil exports to the United States. Economic reality keeps Chavez in check. As long as the oil continues flowing north, Chavez is a minor diplomatic problem, not a concern for the American military. American policy in Venezuela seeks to support local opponents to Chavez who eventually can restore a more democratic and more responsible government in Caracas.

The Frontiers of Freedom

Those who want America to militarily expand the frontiers of freedom should consider how the word "frontier" evokes a certain nuance in the American mind that misrepresents the world as it is today. America's historical sense of frontier was an unsettled or virtually uninhabited space into which people could expand and settle. This idea of frontier, so ingrained in the American spirit, led to a constant expansion into this untamed area that propelled us to the Pacific in the west, the Arctic shores of Alaska in the north, and the Rio Grande in the south. Our continental frontier disappeared in the second half of the nineteenth century, replaced by international boundaries, on the other side of which were settled populations in their own countries. President Taft's decision not to occupy Mexico in the first decade of the twentieth century for all intents and purposes ended the North American frontier. North America's open spaces were now sufficiently populated, no longer there for the taking. Aside from the island acquisitions from the Spanish-American War and Panama, American military expansion was over.

Few Americans prior to World War I would have even considered the possibility that American armies might one day sail across the oceans to participate in foreign wars and subdue vast foreign populations. American participation in the Great War in France was strictly defensive or, perhaps more accurately, counteroffensive. Disillusioned by the political results of the war, America retreated back into our customary continental isolation. Had it not been for World War II, it is likely that America would never have resumed its nineteenth-century pattern of offensive warfare.

The crescendo of righteous vengeance that grew out of the attack on Pearl Harbor animated America's World War II expansion, which culminated in our occupation of two of the world's largest and most modern states, Germany and Japan. But once there, in these two densely populated regions of the planet, our ground forces' limit of advance became a new frontier, not an open space as in the American West but instead a line of separation between two different and populous empires, the American and the Soviet. Our successful occupations were rightly seen as extensions of the frontier of freedom in the sense of an outward expansion of a boundary, not the claiming of an uninhabited expanse.

Worldwide populations have tripled since World War II, disproportionately in the developing world. Extending America's frontiers now means changing the cultural and governmental principles of large populations coming from civilizations very different from our own. Unlike during World War II, some in the global-war-on-terror generation now attempt to move the boundary outward without the strong sense of indignation and national unity that inspired our initial defense in World War II.

Our voluntary neocon expansion into the "gap" is an occupation of a new frontier line, but not one that is uninhabited and, therefore, permissive. Rather, by attempting to occupy and convert the teeming millions on the other side of the planet to American ways of life, we have embarked on a project unlike any ever attempted by this nation: building modern nation-states where they have not previously existed among vast populations who are often indifferent, often hostile, to our efforts to institute American-style governance.

Swimming Against the Tide

In many regions of the "developing" world, the sweeping changes wrought by globalization, our era's dominant trend, work against nation-building, not in favor of it, as weak governments of weak nations find themselves overwhelmed by the growing power and influence of more successful entities. A recent RAND analysis pointed out:

> The state's coercive monopoly—the heart of the Weberian definition of sovereignty—has been eroded dramatically. Across the global south, many states are states in name only, mere juridical fictions that the international community continues to entertain. Challenged from above by globalization (to include structural adjustment programs mandated by international financial institutions), and from below (by non-state forces), the state is finding itself increasingly hollowed out and stripped of its functions. These polities have transferred some of their sovereignty upward, to supranational bodies, and downward, to non-public entities, including private military companies and civil society organizations. When the state fails to provide public goods, self-help structures, including churches and religious sodalities, armed groups, and the informal economy fill the gap.[5]

The trend of weakening central government authority across the "developing world," a post–Cold War phenomenon first popularized by Robert D. Kaplan a decade and a half ago in *The Coming Anarchy*, undermines our offensive nation-building wars in Iraq and Afghanistan as surely as it did in Somalia in the early and mid-1990s. Swimming against the tide of history, our efforts to bring modernity to failed states seems more like a fool's errand than a reasoned national strategy likely to bear results that would justify their costs.

Buffer States

We must also remember that our pushing outward the "frontiers of freedom" will bring us ever closer to the frontiers established by international competitors whom we cannot remove or displace. Eliminating the buffer states between us and the other nuclear powers creates its own dangers. Certain nuclear-armed nations with whom we do not wish to wage war—Russia, China, and even Pakistan—do not welcome American military presence on their frontiers. North Korea exists for a reason: Mao insisted that American troops not be stationed on the Chinese border and was willing to go to war and perpetuate a client state in Pyongyang to prevent it from happening. Russia actively intrigues to roll back the American expansion into the Caucasus and central Asia, in 2008 successfully invading and occupying breakaway regions in Georgia, which the United States wanted to bring into NATO, and in 2009 paying the government of Kyrgyzstan to end American basing rights in that country.

Risking Escalation

But perhaps the greatest reasons why America should rethink its ambitions for global expansion through offensive warfare are not the cost and the futility of the enterprise, nor the saber rattling with our near-peer rivals Russia and China, but the fact that our consequent doctrines—regime change and preemptive attack, as practiced in Iraq in 2003—are impelling unfriendly and unstable nations, namely Iran and North Korea, to accelerate their development of nuclear weapons and delivery systems as their preferred means of defense.

Regimes delegitimized as "evil" and threatened with replacement will consider themselves as threatened by an invasion and act accordingly. The threat that they may miscalculate America's intentions and act in ways that we would consider irrational or provocative is a near certainty. Emerging nuclear powers have a need to manifest their capabilities in ways that can only destabilize important regions of the world and harm our nonnuclear allies. In the worst case, American forces abroad or even America, herself, could be targeted in poorly considered escalatory cycles. As a RAND report for the U.S. Air Force recently concluded:

> U.S. emphasis on the possibility of launching preemptive first strikes against potentially dangerous states armed with nuclear, biological, or chemical weapons may also encourage preemption or other escalation by adversaries seeking to avoid or disrupt an anticipated U.S. attack. Similarly, when regime change becomes a central objective, targets of this policy have weaker incentives for escalatory restraint, especially if the end of their regime is tantamount to their incarceration and execution: Enemies with nothing to lose have nothing to lose in taking extreme risks.[6]

Indeed, our offensive military expansion has not made us more secure, it has only contributed to making Americans and our allies more vulnerable to the fledgling nuclear arsenals of unstable nations. To make matters worse, even in a "limited war," American firepower could topple the governments in either Teheran or Pyongyang, causing nuclear weapons to fall into the hands of internal factions or international terrorists whose motives and intentions may be even more dangerous than those of the governments we destroyed.[7] Unsurprisingly, the report recommends that America abandon its aggressive stance toward these nations and begin to ensure them of our peaceful intentions.

At some point the growing burdens of occupation and the escalating dangers of continued advance will force the culmination of the current American expansion. The neocons contend that we are not yet at that point, if we ever will be. Those of us who are more cautious suggest that we already are. Our twenty-first-century strategic military offensive has cost us dearly and has made the world more dangerous, not safer, for American interests. Perhaps, then, we should consider the alternative, strategic military defense, which characterizes much of America's military tradition. Certainly one can argue that a defensive military policy—attacking only when attacked first—has been far more successful than offensive policy in expanding American influence abroad.

Chapter 13

An Alternative Grand Strategy: Strategic Defense and Limited Wars

N eocons have claimed that a renewed strategic offensive is the best way, perhaps the only way, to promote freedom and American interests at this time. They view the offense as an opportunity created by the collapse of the Soviet Union, our erstwhile superpower rival, and a requirement to defend ourselves against Muslim terrorists operating from sanctuaries in failed states. In their mind, it is necessary to militarily occupy the countries that harbor terrorists, not only to get at the terrorists themselves but to remake these societies along American lines because democracies find ways to live together peacefully.

This view gains force because it combines three enduring American cultural traits: the seeking of opportunity and advantage, the belief that vast conspiracies threaten American liberty and well-being, and the teleological view that Americanism is the natural end state of human development.

To be sure, Americans from the time of the first colonies have viewed themselves, in John Winthrop's famous words, "as a city upon a hill," an example of hope and decency for all to see and emulate.[1] The rights that Americans asserted for themselves in the Declaration of Independence were phrased as the universal rights of man, not just of those living in the thirteen colonies. Surely others would see the virtue of the American system over time. Americans have always considered their love of liberty to be a foundation of their international and military policies.

The Constitution envisions military power to provide for the "common defense" and "secure the blessings of liberty" and later allows Congress to "call forth the militia to execute the laws of the union, suppress insurrections and repel invasions."[2] There is no provision for offensive war; there is no requirement to promote freedom overseas. George Washington, reflecting upon the then-recent war against Britain and the struggle for the new constitution, stated in his first inaugural address in 1789 that "The preservation of the sacred fire of liberty [had been] entrusted to the hands of the American people."[3] Still, Washington rejected alliance with revolutionary France and requested a small standing army of perhaps three thousand men; he thought that a larger army would threaten liberty. The militia, the people in arms, would be the real guarantor of freedom.[4]

At Gettysburg, three-quarters of a century later, Lincoln framed the Civil War as a contest that had to be fought so that "government of the people, by the people, for the people shall not perish from the earth." But Lincoln was talking only of domestic threats to American liberty. Decades before, Lincoln and others had sensed that American freedom was secure from foreign threat within its boundaries. Lincoln told a civic group in 1838, "All the armies of Europe, Asia and Africa combined, with all the treasure of the earth (our own excepted) in their military chest; with a Buonaparte [sic] for a commander, could not by force take a drink from the Ohio, or make a track on the Blue Ridge, in a trial of a thousand years."[5] In 1838 the army totaled only twelve thousand men.[6] But Lincoln, in the broad American tradition, had faith that oceans and militia could defend America against foreign aggression. Only internal American divisions could destroy the American experiment.

With the Civil War won and the Union preserved, Americans sensed that liberty was finally secure in this country. In 1877, even with the Indian wars still in their denouement, the entire army was only twenty-four thousand soldiers.[7] No longer needed to either defend or expand American liberty, the army nearly ceased to exist. Few, if any, thought America should leave its splendid isolation on Winthrop's hill to conquer or defend the gawking populations in the valleys overseas.

America's role in the world was succinctly characterized first by John Quincy Adams as "the well-wisher to the freedom and independence of all . . . the champion and vindicator only of our own."[8]

In 1898 America for the first time since the Mexican War attacked outside its borders to assist the cause of freedom on the nearby island of Cuba and, as a consequence, also took possession of the Philippine Islands in the Far East. The latter war, especially, was considered by many to be a betrayal of American innocence, making us no better than the imperialists and colonialists of Europe. Nevertheless, the army was expanded to 275,000 soldiers to fight these offensive wars before demobilizing to 75,000 men upon the conclusion of hostilities in 1902.[9] Defense of the nation would once again be entrusted to the militia, the seas, and the new Great White Fleet.

Fifteen years later the army would for the first time deploy to Europe to assist in the defense of France. Even though the army expanded to nearly

3.7 million men to fight the war, Wilson's America opted against extending freedom by force after the Central Powers collapsed. Instead of seeing the victory as a historical opportunity to remake the world, Americans rejected the notion of permanent foreign entanglements and military expansion. Rather than embrace Wilsonian internationalism, America returned to Winthrop's hill. American power and wealth, by this point unrivaled, would be used for the further development of America, not military adventurism abroad. Softer forms of power—economic, cultural, and scientific—would spread American influence abroad.

The return to isolationism was reflected in military policy. America largely demobilized its army after World War I but maintained a navy "second to none" in the world. The Washington Naval Treaty, signed in 1922, allowed America a navy equal in tonnage to Great Britain's and two-thirds larger than Japan's. (The other European powers, France and Italy, were allowed navies still smaller than Japan's.) The intent of the treaty was to enable an American strategic defense behind an impenetrable forward line of ships at sea backstopped by the army's coastal artillery that protected America's major ports. That same year, the Congress limited the size of the army to 137,000 officers and men. America's security was provided by the big guns of the navy's battleships and the army's port fortifications. The small ground force that remained after the post-World War I drawdown was dispersed among some forty-five small garrisons, many along the Mexican border.[10]

With the navy as America's first line of defense, the army was essentially placed in mothballs. There it would have stayed had World War II not required America to garrison its Pacific possessions, prepare for the liberation of Europe, and eventually invade and occupy Germany and Japan.

Our traditional pattern of demobilizing the army following war continued after World War II. By 1950, despite America's new overseas commitments in Europe and northern Asia, the U.S Army numbered only 591,000. The U.S. Marine Corps had only 75,000 men. With only one division stationed in Germany and no combat units in Korea, the United States committed virtually none of the army to frontline defense, relying on a 670-ship navy and a 411,000-man air force armed with nuclear weapons to deter aggression and provide for the homeland's strategic defense.[11]

The Korean War compelled President Truman to grow the army to 1.5 million men by 1953, but afterward the army was reduced to less than 900,000, reflecting the perceived need for five U.S. divisions to deter Soviet aggression in Germany and two divisions to react to renewed fighting in Korea. The army ramped up again to 1.5 million men for the Vietnam War.[12] After that conflict, the latter-day Cold War army was reduced to 780,000 men during the late 1970s.

After Desert Storm in 1991, the army gradually fell in strength to 480,000 men, and, as discussed previously, would have been reduced by 50,000 more in the early 2000s, to levels not seen since the 1930s, had the 9/11 attacks not taken place. In 2007, six years into the global war on terror, the Bush administration and Congress finally agreed to increase army strength by 65,000 to a total of 545,000 troops to enable the protracted prosecution of our offensive wars in Iraq and Afghanistan.

This chronicle of the army's ups and downs serves to illustrate some key realities that persist to this day. First of all, the city on the hill—America herself—is not defended by the American army but by the other services, and this has been the case for well over a century. If Lincoln believed in 1838 that no invasion of the United States by a foreign power, or alliance of foreign powers, could succeed, how much less probable would the success of such an attack be today? The U.S. Army currently exists, as it has since 1898, primarily to expand American influence overseas and then defend our newly acquired overseas interests. Second, its defensive missions overseas have been accomplished by relatively small numbers of soldiers compared to the enemy armies they faced. During the height of the Cold War, the five divisions in Europe and two in Korea were only a small fraction of the enemy divisions that opposed them. The success of this relatively small deterrent force is a function of the overwhelming tactical, operational, and strategic firepower, including nuclear weapons, that our air and naval forces can bring to bear on any attacker. Third, given the end of the Cold War—that is, the collapse of the Soviet Union—there exists no military in the world with the capability to mount a credible ground attack on U.S. Army forces that are currently based overseas. The current frontier is entirely secure from conventional ground attack; no nation-state currently has an armed force capable of successfully attacking and defeating U.S. ground forces backed up by U.S. joint firepower from the air force and the navy. Finally, America's tremendous advantage, almost a monopoly, in long-range precision-strike weapons guarantees that any country that attacked U.S. ground forces overseas would, with certainty, find its military and economic infrastructure destroyed courtesy of the U.S. Navy and U.S. Air Force.

Throughout the entire Cold War period and to this day, the U.S. Army and Marine Corps have served as an expeditionary force to defend or expand our overseas frontier. The strategic defense of the United States itself has been secured by the air force and the navy.

Our purpose here is to contemplate the alternative to American strategic military expansion, namely a strategic military defense. What would the future look like if the Department of Defense was actually used primarily for *defense* rather than *offense*? Would American interests suffer? Would we have to abandon our self-made identity as the role model and leader for progress and decency in the world?

I argue that American security and influence in the world would be enhanced by a policy of strategic defense, just as it was during most of the twentieth century. There is simply no need to voluntarily assume the costs and risks of military expansion when our true interests and our historic role can be better served by peaceful pursuits.

Defensive War

Defensive warfare is the measure we take to ensure the survival of our own government and of allied or friendly governments. Defensive measures also seek to protect American citizens and soil from acts of war committed by other nations and organizations—for example, nuclear strikes and terrorist attacks. Deterrence, meaning preparations for defense, is no less a form of defense than is defensive combat. Both seek to preserve existing forms of governance and protect their citizenry. In a larger sense, any action we take in support of a friendly government—economic, humanitarian, diplomatic, or military—helps preserve that government and is therefore defensive.

What the U.S. military calls "stability operations" are in reality deterrent or proactive defensive operations in support of existing regimes, designed to keep opposing forces, whether internal or external, from overthrowing them. All nations enjoy an inherent right to fight defensive wars, prepare for defensive wars, and conduct stability operations within their borders, consistent with international law and custom.

Strategic Defense

Strategic defense is an enduring policy dedicated to preserving the American form of governance in this country and the survival of friendly governments overseas. The policy explicitly recognizes the right of "threat" and rival nations to conduct their own strategic defense. Similar to the American practice during the Cold War, a strategic defense policy requires that nations guarantee not to initiate offensive war or initiate acts of war and that they maintain armed forces structured without a credible offensive war capability. Often, treaties and verifiable inspection regimes are required as confidence-building measures. The "city on the hill" embraced strategic defense internationally. Friend to all and enemy to none, the city sought only to live in peace and freedom.

In truth the U.S. Army conducted a strategic offensive during our continental expansion to the west, and generally paid little regard to the sovereignty or the right to defense of Indian tribes, Hispanic governments, and at times Canada. American expansion into new unsettled territories was seen as our manifest destiny, and the colonists and later the United States constantly initiated offensive wars designed to push the frontier westward. However, with regard to the nations of Europe and Asia (the Philippines and Vietnam excepted), America's historical preference through the end of the twentieth century generally favored strategic defense, fighting only when other nations attacked first.

The rule of law is an essential component of any American defensive strategy. In America the Constitution empowers the judiciary branch to make the executive branch prove probable cause or guilt in a court of law before the police may initiate force against a citizen. The police may act in self-defense or in the exigent defense of citizens without court approval, but they must obtain judicial approval of search warrants, indictments, and convictions in nonexigent cases, meaning where the executive is initiating the current confrontation.

Rule of international law would work similarly. The United States, the executive in the police analogy, would retain the right to self-defense of the United States and our allies by treaty. Other military actions—offensive or limited wars—would be subject to something akin to judicial approval, meaning the permission of other major powers and/or international bodies. Fostering and respecting the international rule of law makes American power more tolerable and productive internationally.

(In contrast, some would suggest that America, having no current peer competitor to balance our military power, should place herself above international law and pursue an aggressive military expansion to seek both national advantage and universal good. Realists would disagree, noting that aggression causes reaction and that coalitions would then rise to oppose and balance American power. Given the practical limits of American expansion discussed previously, such a coalition is likely. Some see it forming already.)

The United Nations Security Council has acted as that judiciary since the World War II victors established it in 1945. As imperfect as the organization is, the UN authorized the Korean War and Operation Desert Storm, America's two counteroffensive wars against naked aggression. The United Nations Security Council has never restricted legitimate American defensive actions. On the other hand, the security council has never authorized any member nation, including the United States, to initiate offensive war.

The U.S government never explicitly asked the security council after the 9/11 attacks for permission to support the Northern Alliance in its overthrow of the Taliban regime, so we can only speculate regarding how that request would have been considered. In any event, American actions in Afghanistan could arguably be considered a "hot pursuit" against a terrorist organization internationally recognized as the guilty party in the attacks in New York and Washington. United Nations Resolution 1378, passed immediately after the fall of Kabul in November 2001, recognized the legitimacy of American military efforts against the Taliban and al-Qaida in Afghanistan.

Although the Bush administration argued that Operation Iraqi Freedom was his way of enforcing UN Resolution 1441, dealing with the inspection of Iraq's alleged nuclear weapons program, the resolution contained no clause allowing the United States to invade Iraq should Saddam Hussein not comply with the inspection regime. Many security council members felt that the United States should have returned to the security council to request permission for the March 2003 invasion of Iraq. If such a mechanism would have forced the American administration to make a stronger case for its offensive action against Iraq, it could hardly be argued that these procedures would have harmed America's military security or reduced our power in the world. As it turned out, Saddam Hussein had no weapons of mass destruction programs and was not a threat to anyone but his own people.

United Nations Resolution 1511, passed by the security council on October 16, 2003, seven months after the invasion, authorized the United States to create "a multinational force under unified command to take all necessary measures to contribute to the maintenance of security and stability in Iraq." The security council, on balance, has recognized that there are legitimate roles that only American military power can play.

Many of the world's nations realize that American use of its military power has, much more often than not, benefited populations overseas. Having a well-meaning nation such as America to maintain peace and security worldwide is not a bad thing in most people's minds, especially when they consider the horrible wars that plagued much of the world before America assumed its leading global role. American military power is not as much unwanted in the world as it is feared. Placing the military more firmly under the rule of law, especially when it comes to offensive war, would relax some of these fears. Just as every citizen deserves to be secure in his home against unreasonable searches, seizures, and arrests, so, too, must nations and their governments feel free from invasion or other offensive action, unless a judiciary rules that their conduct makes them demonstrably criminal or dangerous to others.

Indeed, a recent report from the U.S. Congress underscores the destructive effects of American unilateralism on foreign perceptions of America.[13]

[P]olls conducted by the U.S. Government and respected private firms have revealed a precipitous decline in favorability toward the United States and its foreign policy. The generally positive ratings from the 1950's to 2000 moved to generally negative after 2002. As the very first witness in a 10-hearing series with pollsters and regional analysts told the Subcommittee—"We have never seen numbers this low."[14]

Foreigners generally continue to be favorably impressed by the city on the hill and supportive of American military strength.

* * *

The primary finding from the Subcommittee's series of hearings is that the decline in our standing does not appear to be caused by a rejection of such values as democracy, human rights, tolerance, and freedom of speech. Nor is it a reaction to such facets of American life as a high standard of living, mass culture, and economic opportunity, or to the American people—or even to U.S. military power, so long as it is exercised within the framework of international norms and institutions. All of these were well established and well known prior to 2002 when America's image was at its highest.[15]

Rather, it is America's recent abandonment of its traditional leadership role, no longer forging foreign policies by international consensus or acknowledging restraints on its military power, that propels much of the new anti-Americanism.

It's the unilateralism: A recent pattern of ignoring international consensus, particularly in the application of military power, has led to a great deal of anger and fear of attack. This in turn is transforming disagreement with U.S. policies into a broadening and deepening anti-Americanism, a trend noted by the Government Accountability Office.[16]

According to Dr. Steven Kull, director of the Program on International Policy Attitudes (PIPA) at the University of Maryland, much of the problem results from the Bush administration's decision to enforce UN Resolution 1441 unilaterally:

[T]he complaint is not really that Saddam Hussein was removed. The complaint is that the U.S. did not get UN approval . . . The world is looking for reassurance that the U.S. is constrained by the rules that the U.S. itself has promoted. People are showing more genuine nervousness about whether the U.S. is . . . actually following some new model of its role in the world . . . There was a majority saying that if the United States got UN approval that would make it all right . . . The perception was that the U.S. did not have the right . . . to act preemptively relative to Iraq.[17]

Absent American self-restraint, the world grows fearful of American power.

Among the most shocking data presented to the Subcommittee were polls showing that in many countries, including U.S. allies, majorities believe that a U.S. military attack on their nation is a real possibility. For example, Pew found that 65 percent of people in Turkey, a long-standing U.S. ally, fear that the United States might attack it in a dispute.

This belief is replicated in a number of seemingly unlikely countries, where majorities express fears that the U.S. will use force against them, including in Indonesia (80 percent), Pakistan (71 percent), Lebanon, Jordan, Indonesia, Russia, Nigeria, Morocco, and Kuwait.[18]

Our unconstrained unilateralism has created a worldwide backlash against American foreign and military policy. The considerable stock of American goodwill amassed over the centuries is running low. It will be only a matter of time, should we not change course, when more serious attempts will be made to form coalitions of forces against us, as the realists predict. An American defensive strategy would do much to allay the fears of others and reassert America's moral leadership in the world. To that end, America must pledge to strengthen and support the international mechanisms constraining warfare and accept limit-ations on the legitimate use of force by all nations, especially our own.

Defending America

That there would be a military attack on the United States aimed at imposing a new form of governance on American citizens is, at this time, simply unimaginable. Given the overwhelming superiority of our nation's air and naval forces, not to mention our nuclear weapons, there is no likelihood that the military would have to fight in the defense of our national soil. No nation is currently known to have any ambitions, much less plans and force structure, to invade America. Our declared enemies abroad, al-Qaida and other terrorist groups, have no military ability to change life in America. Their long-term goal of converting America to Islam and bringing it into the Caliphate is a pipe dream, no more likely than the dreams of some Christian proselytizers for the conversion of Islamic nations to Christianity.

With the city on the hill apparently safe from invasion, the only military menace to its citizens is from weapons of mass destruction. Nuclear weapons cannot be wished away. Preventing nuclear detonations on American soil is rightly the number-one priority of the Department of Defense. Many countries possess nuclear weapons and the needed delivery systems: Russia, China, Britain, France, Israel, India, and Pakistan. Others such as North Korea and Iran are seeking nuclear capability. Of the nonallied nations, only Russia and China currently have the combination of weapons and delivery systems to actually reach and destroy targets in America. The Russian capability compares to ours, at least in size. The Chinese strategic nuclear ability is far smaller and much less sophisticated but nevertheless real.

So far, the United States has relied on its enormous and sophisticated nuclear forces, capable of literally destroying any nation on Earth, to deter any ratio-nally calculating government from attacking. American assurances that a strike on the United States, however large or small, would be answered by our own devastating retaliatory strike on the attacker has prevented the employment of nuclear weapons for over half a century. Strong nuclear defense, coupled with our promise not to unleash a nuclear offensive against unfriendly nations, has served the nation well.

Some fear that irrational nations, namely North Korea and Iran, will develop the nuclear warheads and delivery systems needed to attack America. The governments of these charter members of the Bush administration's "Axis of Evil" are viewed as indisposed to make rational calculations regarding war; essentially, they are seen as nation-state suicide bombers seeking nukes. We are told that they will not behave as the Russians and Chinese have for decades but will launch their nuclear-tipped missiles toward the United States as soon as they can produce them, and then spend their last hours, if not mere minutes, of existence waiting for the American retaliatory response.

This unlikely scenario has inspired our current antimissile defense program, the single largest line item in the Pentagon budget. To date, neither North Korea nor Iran has a proven—meaning successfully tested—weapon system able to reach an American city or a country to whom we have guaranteed nuclear protection. Both countries are clearly working on the capability. On the other hand, American progress on strategic missile defense seems to have produced the capability to shoot down incoming warheads in their midcourse trajectories. Iranian chants of "Death to America" reflect immaturity, and perhaps intent, but not yet military capability.

The other "worst case" scenario is that a terrorist group would obtain a weapon of mass destruction, smuggle it into the United States through Mexico or sail it into some U.S. harbor in a shipping container, and detonate the device in a major American city. No terrorist group has anything approaching the technological ability to produce a nuclear bomb, so presumably the bomb would have to be obtained from rogue states that are acquiring nuclear capability, such as North Korea and Iran; from the unsecured arsenals of a disintegrating Pakistan; or from criminal elements in the former Soviet Union.

In response to this threat, the 9/11 Commission recommended, and many agencies of the U.S. government requested, more stringent inspection regimes of all cargoes destined for American ports or points of entry. One system suggested by the U.S. Coast Guard would require U.S. officials to inspect and seal all shipping containers destined for America at their overseas ports of origin. Shipping manifests could be audited by U.S. counterterrorism agencies as the ships were being loaded overseas. Coast Guard planes would then fly over the container ships midocean and electronically query all the shipping seals, check the results against the manifest, and thereby ensure that no uninspected containers entered U.S. ports. Presumably, this would greatly reduce the risk of terrorists sneaking a nuclear weapon into America.

The cost of such a regime would be a tiny fraction of the value of the goods shipped, raising, for instance, the purchase price of a Chinese-made television set at an American Wal-Mart by perhaps a dollar. But the Bush administration did not require or fund such inspection systems, and currently a large percentage of the containers entering the country are not inspected. Clearly, the government would be more concerned if it thought that terrorist-smuggled nuclear devices were actually a threat to the city on the hill.

But what if even the worst happens? Let us suppose that either North Korea or Iran managed to assemble a usable nuclear weapon, and then figured out how to build an intercontinental missile to propel it, then secretly launched it through America's missile defenses, and the bomb flew true and detonated over an American city. Or let us suppose that some undetected terrorist plot managed to detonate a nuclear device within the American homeland. Would it mean, as some pundits suggest, the destruction of America and our way of life? Would freedom and prosperity no longer bless the land? Would the American experiment end?

Clearly the answer is no. Tens of thousands of Japanese died in the atomic bombings of Hiroshima and Nagasaki in August 1945, but the Japanese nation did not die. As awful as the loss of life in the bombings was, the nation survived and people throughout the rest of Japan went on with their lives. Hiroshima and Nagasaki were rebuilt. Japan moved on to even greater prosperity and freedom than it enjoyed before the bombings.

I certainly doubt that Americans would prove themselves less resilient than the Japanese. Every death is regrettable. Tens of thousands of deaths would be a heart-rending national tragedy. But it would not be fatal to the nation. As we must minimize the risk of such an attack, we must also correctly assess its likely impact. Those who claim that we are in a life-or-death struggle with our current enemies are wrong. The Russian nuclear forces excluded, there currently is no existential military threat to America itself.

Defending Our Friends and Allies

Our primary defensive concern overseas is that one of our friends or allies such as Israel, Kuwait, South Korea, or the newly independent countries of Eastern Europe would be attacked or threatened. First of all, our principal allies in NATO—as well as Israel, Japan, South Korea, and Taiwan—have more than enough conventional forces to repel any likely aggression. Russia no longer has an army capable of threatening NATO with invasion; the Israelis are by far the most powerful military force in their region; North Korea's army is decrepit; and China does not have the amphibious capability needed to invade Taiwan. Second, American military power backstops our allies and provides a forward defense that no nation is willing to test. Since World War II, no government defended by U.S. troops has been invaded by another nation's army. (Korea and Kuwait were attacked before America's military commitment had been made clear by the stationing of army combat brigades.)

Indeed, the tremendous fighting capability of U.S. ground forces, backed as they are by joint fires and nuclear weapons, has proven to be a decisive defensive deterrent against conventional attack. Our alliance system with other economically developed, democratic allies has proven strong enough to deter all potential aggressors. Over the past decade, we have withdrawn most of our forward defense brigades from Korea and Europe with no ill effect. For the decade between Desert Storm and OIF, U.S. brigade combat teams rotating through Kuwait were sufficient to defend our friends in the Gulf. The U.S. Army, supported as it is by lavish joint fires, needs relatively few brigade combat teams for forward defensive purposes. Currently the army stations only two brigade combat teams in Europe and a single brigade combat team in Korea; the remaining seventy-odd brigade combat teams are either fighting the offensive wars in Iraq and Afghanistan or are in reserve at army posts in the United States.

The main military threat to our allies is, again, the nuclear weapons programs of North Korea and Iran. Clearly any strike by North Korea against Japan or South Korea would require an American retaliatory response, since we have pledged to secure these countries under the umbrella of our nuclear deterrence as a quid pro quo for their remaining nonnuclear nations in accordance with the Nonproliferation Treaty.

A strike by Iran against Israel is a different matter. In the 1970s Israel developed its own "undeclared" nuclear forces and is now armed with about two hundred warheads, according to knowledgeable sources. Israel chose to develop its own nuclear deterrent and retaliatory capability outside of American control and in violation of the Nonproliferation Treaty. Israel has also demonstrated its willingness to preemptively attack Arab nuclear facilities and is no doubt studying the Iranian program, especially in light of the rhetoric coming from Teheran.

As regrettable as any nuclear strike anywhere in the world would be, attacks between regional enemies overseas would not pose an existential risk to the city on the hill. Even as we use our good offices to try to keep peace in the valley, Americans need not make themselves responsible for all the bad things that happen there, especially those things that are a consequence of willful disregard of international law or treaty obligations. Nor do we have a duty to foresee and prevent every calamity. Other nations must bear their own risks and suffer the consequences of the decisions they make. We need not make ourselves the world's nanny.

A smaller concern is unrest or insurgency within friendly, less-developed Third-World countries struggling with democracy and afflicted by poverty and internal divisions. Winning these wars is generally not vital to America's survival or prosperity, so their importance must be kept in perspective.

American military, economic, and governmental assistance—rather than U.S. brigade combat teams—appropriately serves as the backbone of our defensive efforts in these developing regions. Regarding the military aspects, the influential author of *Learning to Eat Soup with a Knife*, John A. Nagl, called for a twenty-thousand-man permanent advisor corps to improve these nations' abilities to defend themselves, as well as increase the bonds between their militaries and ours, which is clearly preferable to deploying American combat forces for their defense. Since nearly all of this advisor corps would consist of mid-career or senior officers and noncommissioned officers, not privates and lieutenants, these twenty thousand advisor slots compete with the army's new brigade combat teams for valuable personnel and force structure. However, significant manpower savings could be harvested from the inefficient brigade combat team designs.[19] And successful advising efforts would reduce the need for American brigade combat teams to conduct defensive, counterinsurgency, and stability operations themselves. A large corps of military advisors with operational assignments in various regions and countries, who possess language and cultural skills, who have detailed local knowledge, and who have made investments in personal relationships cannot help but make common defense easier for both America and the nations immediately involved.

Defensive counterinsurgency, meaning efforts to defend a legitimately empowered, popular form of government, must be distinguished from offensive counterinsurgency, which aims to eliminate resistance against an occupier.[20] In defensive counterinsurgency, the government generally has the ability to broaden its political base to co-opt insurgent demands and defuse the crisis. Occupiers, because they are foreigners who have usurped sovereignty, rarely have the flexibility to negotiate the terms of their governance with the insurgent and by their very presence inflame insurgent passions. Offensive counterinsurgency is often a bloodbath, as seen in Vietnam, Algeria, Soviet-controlled Afghanistan, Chechnya, and Iraq.

Defensive counterinsurgency can be much more contained. All defensive counterinsurgencies are different, but none to date has required a large commitment of American ground forces. Small amounts of American military, economic, and political assistance have proven effective and likely will continue to be effective. Military-to-military contact—for example, combined exercises, deployments, foreign military sales and aid, providing international military students professional training here in the United States—and a host of other programs create military relationships that serve stability.

Limited Wars to Enforce International Order

Limited war involves those military measures, short of regime change, taken to force a government to change its policies or actions. The essence of limited war is to inflict military pain, or threaten to inflict military pain, to change the behavior of a nation, not its governance. Diplomatically identifying and isolating the behaviors in question is a necessary antecedent to limited war. Maintaining focus on those behaviors is necessary for prosecuting limited war. Also critically important, belligerents must reduce the likelihood of unwanted escalation by clearly enunciating the measures they are willing to take and thresholds that cannot be safely crossed by their adversaries. In order to achieve closure to the conflict, all the parties involved must be allowed to resume normal relations once they have agreed to the resolution of the behavior. Unless it is believed that these conditions can be achieved, launching a limited war is folly.

Our war with Iraq from Desert Shield to the beginning of OIF was primarily a limited war, principally designed to restore the pre-invasion border with Kuwait, protect the Kurds, prevent Saddam's reacquisition of weapons of mass destruction, and ensure that Iraq would never again become an offensive threat in the region. Regime change, the U.S. overthrow of the Hussein dictatorship, came later.

The nation launching the limited war must have reason to believe that limited offensive action will force favorable resolution of the issue in question without the conflict escalating to the level of regime change. The problem is that the defender may not agree to limit the war. For instance, the Japanese attack at Pearl Harbor was not aimed at destroying the U.S. government in Washington, D.C., but to force the Americans to accede to Japanese expansion in the western Pacific. However, the American reaction to Japan's limited offensive—total, unconditional war—forced Japan into a desperate war of attrition that ended with the American occupation of Japan. Imperial Japan's attacking of a much stronger country was a high-risk gamble that ultimately led to its defeat and destruction. Limiting a war is always problematic. The attacker may want limited war, but the defender will always decide when the war will end. An attacker always risks opening a Pandora's box.

The United States today is not Imperial Japan of 1941. Our military problem is not how to attack a stronger nation but rather how to focus our military power in countries far weaker militarily and achieve worthwhile results. Clearly we have created the means to act, effectively or otherwise, but are unclear as to how limited military operations can achieve our political aims.

When it comes to imposition of will, we have difficulty determining how much success is enough. Incomplete victories in our limited wars have served only to whet the appetite for more complete victories. We sometimes fail to constrain our impulse to enlarge our conflicts into ever-greater struggles, which eventually tend to lead us into an offensive war that we are unprepared to win. No doubt, the bellicose statements made by opposing government leaders taunt us; Iranian president Mahmud Ahmedinejad and Iraq's Saddam Hussein before him inflamed American animus, which led to hawkish calls for offensive war.

Our unfortunate pattern is to escalate limited conflicts into wars of occupation and, consequently, assume attrition and other costs for which the political base supporting the war is unprepared. In America's recent experience, our initial limited war with Iraq, Desert Shield/Desert Storm, has morphed into a protracted and escalating struggle known as Operation Iraqi Freedom, which has no foreseeable end. But before OIF was Vietnam, before that Korea, before that the Philippines, before that Mexico, and before that the invasion of Canada in the War of 1812, all of them testaments to America's tendency to escalate limited conflicts well beyond their original boundaries. Fortunately, some of our most extreme proposed escalations—for example, the calls to liberate or destroy Red China during the Korean War or to invade North Vietnam— have been wisely resisted by political leaders.

Americans also have an unfortunate mental ability to magnify local wars into existential threats. We have imagined the Japanese invading California, Communists tipping dominoes that would fall until they reached our borders, the Sandinistas threatening Texas, Islamic terrorists destroying the American way of life, and, with regard to Iraq, our failure to defeat "them" there inevitably leading to "them" bringing the war here.

Our difficulties in reasonably assessing and bounding the motives and reach of our opponents present a major liability in prosecuting limited wars. Given our superpower military and economic strength, the vastness of our country and population, and all of our allies and friends worldwide, these fantasies of existential threat are more expressions of their proponents' psychological make-up than they are reflections of some concrete reality. Certainly, terrorists have demonstrated that they can create casualties on American or European soil, but there is no evidence to suggest that they can bring down a government or change a way of life. Certainly, the remaining members of the "Axis of Evil," plus a few more petro-dictators who have emerged since 2002, can test us regionally, but none threatens American existence. In reality, we are at this moment much more existentially threatening to them than they ever could be to us.

Limiting War

America has no real theory of limited war because we have rarely ever sought to fight limited wars. Ironically, Clausewitz, who I have lampooned as naive regarding offensive war, is an excellent source of philosophy regarding limited war.

To Clausewitz, a child of the European dynastic system, wars were fought between the established, interrelated monarchies over relatively minor matters: border provinces, alliances, overseas possessions, and the like. War was an extension and escalation of politics designed to force the other monarch to bend to one's will and concede the question at issue. Never was war aimed at the overthrow of a reigning monarch, an act that could collapse the legitimacy of the dynastic system. If all monarchs in all countries ruled by the grace of God, as they told the masses and themselves, then the positions of all monarchs must be respected by their brother monarchs, lest confidence in the entire scheme dissolve.

The dynastic system is an essential background to Clausewitz's observation that "war is thus an act of force to compel our enemy to do our will."[21] Importantly, enemies are not eliminated but rather forced to act in agreeable ways. Given the limited nature of conflict, Clausewitz later stated that his task in *On War* was to combine the unlimited violence of battle at the tactical level with the assumed limits on warfare that existed at the strategic level. In war, the "blind natural forces" of "primordial violence, hatred, and enmity" must be subordinated to policy, which is "subject to reason alone."[22] Hence, Clausewitz's famous dictum, "[W]ar should never be thought of as something autonomous but always as an instrument of policy . . . the first, the supreme, and most far reaching judgment that the statesman and commander have to make is to establish . . . the kind of war on which they are embarking; neither mistaking it for, nor trying to turn it into, something that is alien to its nature."[23] In other words, don't let the emotions of war overwhelm your rational calculations and well-considered motivations.

Those who would have America reflexively use unlimited military force to solve limited problems tend to undervalue Clausewitz's advice. For instance, those who have advocated using military force against Iran tend to be very unclear regarding the kind of war they would wage or what the achievable end state of the war would be. Consequently, we cannot judge whether the results of the war would justify the costs and risks. In the same vein, the neocons claim that America's only real option concerning Radical Islamists is to invade their sanctuaries in the "gap" and establish pro-American governments capable of keeping them in check. In the neocons' view of the world, America must respond to terrorist attacks, or threats of such attacks, with wars of occupation that we initiate and then become compelled to win. Terrorist provocation unleashes American escalation in the form of an invasion and nation-building exercise in some failed or failing state. Consequently, America voluntarily incurs tremendous military and financial costs for a great many years with no assurance of "victory," or certainty that "victory" will result in benefits for the U.S. citizenry at least equal to their costs. By reflexively escalating each terrorist provocation to unlimited war—for example, the global war on terror—we actually weaken America and play into the hands of our enemies. The notion that America could defend itself with actual defensive wars does not appeal to neocon ambitions, which tend toward aggression.

All too often lately, the emotional energy urging something be done is not matched by a rational calculus of policy objectives and military means, at least in the public sphere. Neither does one often find a discussion of the countermeasures the enemy can take or the potential reactions of third parties. The underlying assumption is that we will simply turn up the military heat and escalate the violence until we achieve victory.

The problem is that we can commit the nation to a costly war over a minor interest, or suffer diplomatic or economic reversals that would actually worsen the situation. We must also consider that the nation we attack today will not go away tomorrow. We may perceive our actions as being limited and we may soon forget the event, but other nations have longer memories.

The Prospects of Strategic Defense

A defensive grand strategy would require that America assure the world that we will not initiate offensive wars, or even limited wars, without a declaration of war from the Congress and preauthorization from an international body such as the United Nations Security Council. (Obviously, we would retain the right to respond unilaterally if attacked first.) We would openly state that we would use American offensive military power only to enforce international norms, agreements, and resolutions, and only with the consent of the appropriate international bodies. The model would be our actions against Iraq from 1990 to 2002, when American military power, with United Nations approval, expelled Saddam Hussein from Kuwait, enforced no-fly zones and other military restrictions on Iraq, protected the Kurds in the north, enforced economic sanctions, and eliminated Iraq's weapons of mass destruction programs. Another example would be U.S. support of NATO, the European Union, and the United Nations to defeat Serbian aggression against the other republics of the former Yugoslavia in the 1990s. It would be difficult to argue that gaining multinational approval for U.S. defensive or limited military action has harmed American security or the interests of international peace.

By all evidence, a defensive policy appears to be the surest way to achieve our global goals. The two great expansions of world freedom in the last hundred years, the first at the end of World War II and the second at the end of the Cold War, came after decades of American strategic military defense, not strategic military offense. The revolutionary power of America has not been its military, but its example of democracy. The new cities on the hill we nurtured in Central Europe and northeast Asia after World War II brought down the iron curtain and opened up China more surely than offensive American military power ever could have. American expansion at the end of the Cold War was mostly economic, diplomatic, and cultural—freedom on the march. American military power moved into Eastern Europe and central Asia not as invaders but rather as the invited guests of self-liberating local governments, now anxious to defend their newfound sovereignty and build their own cities on the hill. Attraction, not compulsion, extended America's influence.

There is no reason to believe that the next wave of American expansion, perhaps a decade or two in the future, could not be the same. Given time and enlightened policies, including American security guarantees, these new cities on the frontier will grow and flourish, and will in turn become beacons to those not yet living in freedom. Dictatorships will find their ambitions thwarted by American defensive efforts and their power undermined by the companion attractions of liberty and prosperity. But the "attraction" process takes both time and confidence. Those consumed by impatience and fear are quick to turn to the military instrument, but in so doing they betray the vision of America's founders.

Strategic defense is a feasible option against Islamic terrorists. Although we cannot prevent every terror attack against America at home and our allies abroad, we can limit the scale and frequency of such attacks to acceptable limits, much like law enforcement agencies cannot stop all crimes from happening but can reduce their incidence sufficiently so that civil society can flourish. We can "manage" the terrorist challenge using internationally supported law enforcement techniques rather than trying to "solve" it militarily and, more and more often, unilaterally. We can make the world safer for Americans and less rewarding for terrorists without resorting to invasions, occupations, and unrealistic nation-building efforts. Noting that "defeating the global jihad will require a long-term strategy designed to cultivate and empower moderate elements in the Muslim world while simultaneously defending them and Western interests from radical Islamists," the previously cited RAND study suggested the following lines of effort:

- *Vigilance, prevention, and defense.* Whether better defending key targets actually deters terrorists from attempting attacks or simply renders their efforts less effective is largely transparent from a strategic point of view. Either way, the only way to reliably constrain Islamists from escalating the global jihad is to deny them the ability to do so wherever possible. The United States and its partners in the struggle against terrorism should continue efforts to refine domestic and international security procedures, share intelligence, and develop new methods and technologies to detect, track, and seize terrorist suspects, weapons, and resources before they can be used against friendly targets.

- *Concentrating visible efforts in the diplomatic and judicial arenas.* Given that escalation generally serves the long-term objectives of global jihadists, the United States and its partners should avoid militarizing the conflict to the maximum extent possible. Operations against terrorists, particularly those visible to international audiences, should emphasize capture and trial rather than military strikes and invasions. Coercive diplomacy may be unavoidable when confronting states that support terrorism, but if conventional military force must be used, it should be used judiciously and with discretion.

- *Entrusting security operations in the Muslim world to local authorities.* Given the history of colonialism and the sense of Western victimization prevalent among many citizens in the Muslim world, Western powers should, to the greatest extent possible, leave military and law enforcement operations in those regions to local authorities. The West should keep its military footprints small to avoid projecting an air of occupation and neocolonialism.

- *Cooperation and foreign assistance.* Keeping the conventional footprint small does not mean noninvolvement. Entrusting security operations to local authorities requires helping states develop the capacity to perform those functions effectively and in ways that do not alienate their own citizens. The United States should cultivate cooperative relationships in the areas of intelligence, law enforcement, and military training and assistance. U.S. intelligence and military specialists need to work cooperatively with their regional counterparts to train and advise their forces and to develop intelligence resources to support U.S. security requirements. The initial phase of Operation Enduring Freedom and the CIA's cooperative efforts with Albanian authorities culminating in the 1998 Tirana raids stand as positive examples of effective operations in the struggle against militant Islam.

- *Restraint and discretion in force employment.* Despite the imperative to minimize conventional military operations, strategic considerations sometimes demand the deployment of conventional forces. When such forces are needed, they should be used, primarily, to provide security to citizens who would otherwise be exposed to extremist violence. Nevertheless, the use of conventional military force in more traditional combat roles may be desirable when opportunities to preempt terrorist attacks arise and may be altogether unavoidable in cases of self-defense. Even then, however, managing escalation requires keeping the use of force to a minimum. This is particularly important in and around cities. Using conventional military force in populated areas, even when employing precision-guided munitions to minimize collateral damage, lends credibility to extremist propaganda and antagonizes noncombatants, strengthening the jihadist movement. When terrorist or insurgent targets of opportunity arise, every effort should be made to strike them away from cities. The CIA's November 2002 employment of a Hellfire missile fired from a Predator drone, killing a high-ranking al Qaeda leader and several other operatives in rural Yemen, offers a positive example of the discrete use of force to achieve an important tactical objective.[24]

How much better America would be had we not responded to the provocation of 9/11 by invading and occupying both Afghanistan and Iraq with our entire available armed forces, but rather had focused our efforts against al-Qaida, the sole source of the 9/11 attack. One could easily imagine a more concentrated raid capturing or killing Osama bin Laden and other al-Qaida leaders at Tora Bora in December 2001 rather than letting them escape into Pakistan. Losing focus on bin Laden to free military assets to expand the global war on terror into Iraq was a crucial early mistake. With bin Laden and his henchmen captured or dead, the likely result would have been early closure on the 9/11 incident.

Having already avenged the 9/11 attacks by the end of 2001, it is doubtful that American policymakers could have rallied public opinion for an invasion of Iraq in 2003. Thousands of coalition soldiers and hundreds of thousands of Iraqis would probably be alive and uninjured today had that war never occurred. Hundreds of billions of dollars worth of debt, the result of the war in Iraq, would not be passed down to the youth of America. Modest programs, commensurate with our modest interests, could have bolstered the post-Taliban government in Afghanistan without trying to impose unwanted occupation and Westernization. Rather than rallying Muslims to repel our aggression, we could have assured them of our moderate aims and the appropriateness of our military method. A reasoned and appropriately defensive response to 9/11, ably executed, would have left America safer and cost America far less than the offensive course on which we embarked and in which we now feel trapped.

It is not too late to demilitarize our war with terrorism. Nor is it too late to realize that we cannot install by force unwanted governance in the Muslim world. We need to realize that our main security interest is to not be attacked by Islamic nations or terrorists operating from them. The form that governance takes in these areas is not a vital American national interest.

This does not mean America shouldn't peacefully promote moderate government as opportunities arise. It does mean that we should pledge not to commit acts of war in the region unless we are attacked and that our military response will be proportional to the attack. We should also put certain foreign governmental authorities on notice that they will be held responsible for terrorist groups operating from their territories and provide them the support needed to ensure that they can police terrorist elements.

Meanwhile, America and its allies should strengthen defenses against terrorist attacks at home and vital interests abroad. Though this is not a 100 percent solution to terrorist attacks, nothing more can be practically achieved without sacrificing more important policy objectives, such as financial solvency, our good name in the world, and peace among nuclear powers.

Chapter 14

The Roman Dilemma: Why We Must Choose between Republic and Empire

Of all the enemies of public liberty, war is perhaps the most to be dreaded, because it comprises and develops the germ of every other. War is the parent of armies. From these proceed debt and taxes. And armies, debts and taxes are the known instruments for bringing the many under the domination of the few . . . No nation could preserve its freedom in the midst of continual warfare.

—*James Madison, 1795*

[T]his country must go on the offensive and stay on the offensive.

—*George Bush, 2004*

English historian Edward Gibbon published the first volume of his magnum opus, *The History of the Decline and Fall of the Roman Empire*, in 1776, the same year the Americans wrote their Declaration of Independence. Gibbon's book was an immediate bestseller and went through six printings, unprecedented at the time. Gibbon published the second and third volumes in 1781, the year of the Battle of Yorktown, and the final three volumes between 1788 and 1789, coinciding with the United States' first year under its new Constitution. Gibbon's work became a mandatory read throughout the English-speaking world, including America.

The rediscovery of the ancient world during the Enlightenment inspired much of America's approach to governance. Mindful that the United States was embarking on the first large-scale experiment in popular government since the ancient democratic city-states of Greece and the Roman Republic, American leaders were eager to emulate the goodness of these examples from the past and to avoid the pitfalls that led to the decline of the classical world into despotism.

From the architecture of the new capitol to the imagery and language on the Great Seal, Americans invented themselves along classical lines. We would be the new Rome—the Republic, not the Empire. The Founding Fathers admired little in Rome's caesars, just as they rejected their European descendents, the kaisers and tsars and kings. The nation's capital would be built around the legislature, the House of Representatives, and the Senate. The people would be the sovereigns of the new republic. And freedom would endure as long as the American people, unlike their Roman predecessors, guarded their liberties.

Mindful of the Roman and European precedent, the Founding Fathers recognized the danger posed by standing professional armies and instead put their faith in the militia. As President Thomas Jefferson wrote in his June 18, 1813, letter to future president James Monroe:

> Every citizen should be a soldier. This was the case with the Greeks and Romans, and must be that of every free state . . . The Greeks and Romans had no standing armies, yet they defended themselves. The Greeks by their laws, and the Romans by the spirit of their people, took care to put into the hands of their rulers no such engine of oppression as a standing army. Their system was to make every man a soldier, and oblige him to repair to the standard of his country whenever that was reared. This made them invincible; and the same remedy will make us so.

Jefferson was writing of the Roman Republic, not the Roman Empire of the caesars.

Before there was the Roman Empire, there existed a republic. The Roman Republic began in 509 BCE with the overthrow of a small monarchy that governed a few tribes in what is now central Italy. The new republic created a constitution that conferred the majority of power to the patrician Senate and the remaining power in the more broadly selected plebian legislative assemblies.

From the founding of the republic, complicated systems of checks and balances diffused power throughout the Roman system. Like modern parliamentary governments, the executives were selected by the Senate and the assemblies but served only brief terms with no possibility of succession. The Senate and, to a lesser extent, the popular assemblies held the real power. For instance, the people would elect up to six commanders for each Roman legion, each in charge on alternate days, to ensure republican control over military matters.[1]

Roman legions in the republican era were not standing or professional armies. For the first four hundred years of the republic, legions were raised from the citizenry when the Senate determined that the republic was threatened or, more likely, needed to launch a military offensive. Each province of the republic would be levied for its quota of soldiers. Soldiers had to be men from the middle or upper classes of the society and had to supply their own armaments. These soldiers would be released back to their homes and careers once the war ended. In this manner Rome not only raised the manpower for war against its enemies but defended its own liberties, the belief being that free men who enjoyed the benefits of Roman society would not attack the republic that enabled both their freedom and prosperity. Moreover, Romans understood that the republic had the right to expect their service as a duty of citizenship.[2] (The American militia system and the volunteer levies of the Civil War were latter-day equivalents of the Roman system.) Though the legions of the republic, hastily assembled and trained as they often were, occasionally lost battles, they never lost wars.

Rome's republican prosperity created a rapidly expanding population, which, coupled with the Romans' cultural legacies of tenacity and martial spirit, made the republic a military powerhouse.[3] Consequently, Rome withstood test after military test and expanded at the expense of its defeated neighbors. In episodic warfare over a period of centuries, Rome eventually defeated both Carthage and Greece, its major rivals in the Mediterranean, and pushed the frontiers of Rome across the Alps into Gaul and Spain to the west, into Macedonia and Anatolia to the east, and into North Africa to the south. By 100 BCE Rome governed more than 1.2 million square kilometers of territory, more than three times the landmass of modern Italy. In its enormously successful republican years, Rome had grown a hundredfold to become the unrivaled military power in its Mediterranean world; no military force on Earth threatened the Roman homeland in the closing decades of the second century BCE.[4]

The decline of the Roman Republic coincided with the rise of the empire.[5] The legions of free men who had served Rome for centuries were inappropriate for the chronic warfare at Rome's distant frontiers in Gaul, Spain, Greece, Africa, and Asia. Free men from the propertied middle and upper classes found little reason to fight these far-off wars against scarcely known enemies who posed no immediate threat to Rome proper. Worse, no man of means could afford to leave his farm or business for the two to three years necessary to form and train a legion, march it to the frontier, fight the war, and march it back home. Bankruptcy and broken families too often greeted Rome's foot soldiers on their homecoming. Consequently, the republic found it increasingly impossible to raise the necessary numbers of citizen soldiers needed to defend and expand its vast empire.[6]

Responding to this crisis in military manpower, Rome enacted the Marian Reforms in 107 BCE, professionalizing the army and creating standardized permanent legions.[7] Soldiers were recruited from all classes for the first time, and the poor flooded into the legions. The term of enlistment was set at sixteen years, after which the soldier could retire comfortably on arable land provided to him by the legion commander he served.[8] In this scheme, the land would be acquired by future conquests, building into the legions a self-interest in military expansion. Later, the legions would begin to accept non-Romans for service under the same contractual formula.

From a purely military standpoint, the new professional legions were a great success. With soldiers staying longer, they could be better drilled and disciplined. Legion commanders got to know their men and vice versa, creating strong loyalties that carried the day on many difficult battlefields. Standardization led to improvements in armament and more consistent battlefield performance across the Roman army.

All in all, the new legions constituted a much more effective fighting force than did the hastily assembled legions of the old Roman tradition.[9] Their effect was immediately seen all along the frontier line; in the fifty years following the Marian Reforms, Rome would expand the area under its control by an additional 750,000 square kilometers, two more Italys. Greater conquests would follow in the first century CE until Rome eventually controlled, at its height, over 5 million square kilometers of Europe, Africa, and Asia.

However successful the professional legions were at creating empire, they proved to be the poison that destroyed Rome's republican form of government. Barely one enlistment period into the Marian Reforms, these new legions were attacking the republican government in Rome proper. The proximate cause of the first military coup d'état in 88 BCE was a political struggle between certain Senate factions and the popular assemblies over whether the powerful counsel and general, Lucius Cornelius Sulla, would lead his legions on new conquests in Greece and Asia Minor.

In an event unprecedented in Rome's republican history, Sulla took matters in his own hands and marched his most loyal veteran legions—some did mutiny in protest—north into the capital and purged his enemies in the Senate, weakened the powers of the assemblies who didn't support him, and put his political allies in positions of power.[10] Feeling that his back was secure, Sulla then sailed off to war and victory.

Returning to Rome in 83 BCE, Sulla was determined to finish the work he had begun four years prior, but his opponents by now had loyal legions of their own. Culminating the bitter Roman civil war that ran from 82 to 80 BCE, Sulla and his victorious antirepublican legions captured and sacked Rome itself. He had the Senate declare him Rome's dictator, not for the traditional six-month period required in times of emergency but indefinitely, and proceeded to execute thousands of the nobles and other citizens who had opposed him. Sulla's purges went on for months. Many of those spared were proscribed from ever again holding positions of power. Sulla died in 78 BCE at the age of sixty.[11]

Though many of the political institutions that Sulla destroyed were in some measure restored after his death, the republic never fully recovered. One of those spared in Sulla's purges was the young Julius Caesar, who, mindful of Sulla's example, would march his powerful legions, fresh from victory in Gaul, across the Rubicon in 49 BCE to initiate a new Roman civil war.[12] Declared by the Senate to be Rome's perpetual dictator in 44 BCE, Julius was soon assassinated by his republican enemies.[13]

However, Octavian, his nephew and adopted heir, would continue Julius' work, using the legions under his command to consolidate power in his own hands and destroy, in yet another civil war, the legions loyal to his uncle's republican enemies. In 27 BCE the victorious Octavian had the Senate proclaim him the first Roman emperor, Augustus Caesar. In that capacity he became the commander of all legions, wherever they might be posted. Legion commanders henceforth reported directly to him, not the Senate.[14]

Throughout his forty-one-year reign, Octavian consolidated Rome's power in his own hands and paved the way for dynastic succession. Though the Senate continued, it became largely a consultative body to the emperor. Rome's other republican institutions, though often maintained officially, were similarly made to conform to Octavian's enormous imperial powers.[15]

Gone, too, were many of the freedoms that Rome's citizens had traditionally enjoyed. Tacitus, the Roman historian, called Rome's condition under the new caesar "slavery."[16] By the time of his death in 14 CE, Octavian had ended the Roman Republic and replaced it with the Roman Empire, a monarchy under his family's control.

It would be wrong to say that the Roman Empire, having destroyed its republican antecedent, immediately fell into decline. Many historians, Gibbon among them, argue that the empire reached its greatest extent and wealth in the second century CE. Rome's professional armies had a knack for conquering nations and extracting wealth from them, at least for a while. But the results of the Roman experiment speak for themselves. Over time, as Gibbon chronicles, Rome's republican virtues were replaced by the vices of empire. The political base of the empire, increasingly corrupt and self-absorbed, narrowed, while the now-disenfranchised citizenry invested themselves in other pursuits, among them Christianity.

Eventually the cost of empire proved too great and the barbarian armies on the frontier too numerous. Rome began to shrink and then fracture. The enduring cultural strengths of the Roman Republic were only slowly exhausted by the Roman Empire. But over the centuries Roman society had gradually been forced to abandon its republican identity to the schemes of the despotic and powerful. In the end, Rome fell because its citizens would not defend the empire.

The lessons of Rome still echo across the millennia. To the Founding Fathers, the lessons most important to their endeavor seemed to be the creation of a constituted government with enduring checks and balances and the need to avoid professional standing armies. In modern America, we should be more concerned with the pernicious effects of empire on the health of the republic.

America, in the broad view, is following the same trajectory as did Rome; both started as small and virtuous republics, tapping their own popular energy to grow and prosper. Militarily, both fought their attackers with armies hastily raised from among the people, eventually defeating their major military rivals in a series of world wars. Militarily superior, they both continued to expand their frontiers outward until, finally, both decided that professional armies had become necessary to maintain and further expand the empire. Both then strengthened the powers of the executive branch at the expense of the legislative branch as the demands of the military and overseas adventures assumed greater importance in the affairs of state, and more populist local concerns assumed lesser importance.

Past this point, one can only speculate about the future of America. There is nothing automatic about history. America does not have to follow Rome into despotism, overextension, and collapse. We can make choices other than the ones Rome made, or had made for it, and ensure that our republic will exist not in the wistful reminiscences of old men but in the daily lives of our posterity.

What must be done to defend the American Republic?

First of all, we must realize that our imperial ambitions to Americanize the world by military force are unnecessary to our own defense and will only serve to mire the nation in constant warfare, which will be our undoing as a free and prosperous country. Empire becomes a trap from which there is no escape. Expanding the frontier by military conquest does not end war but only moves the war to a new frontier against new opponents. Defeating one enemy only exposes another, which is then viewed as a threat to be defeated. The process repeats itself with no apparent end. Rome "defended" itself all the way to the Atlantic coasts of Spain and France, across the Channel to England, into Germany in the north and to the Danube in the east, across Anatolia into Persia and Palestine in the southeast, and south down the Nile valley in Egypt. Never was there an "enough." Rome's offensive spirit always pushed the empire onward, convinced that attacking the enemy on the frontier was better than defending against him or just letting him be. Enemy resistance to Roman expansion, even the defeat and destruction of entire Roman legions, only steeled Rome's determination to send even more legions to the embattled frontier to finish the conquest.

The current American proponents of military expansion, like the Romans before them, see no limit to our empire, only an embattled frontier needing more forces. No victory brings peace, just a new, more distant battle line. As we expand, we make friends and enemies we never knew existed, and these new friends and enemies become the reason for yet another war.

Most Americans have difficulty locating our new imperial battlegrounds on a world map. "Where is Kurdistan, anyway?" Still fewer understand the subtleties of the wars we are waging. How many understood the significant differences between Sunni and Shiite Muslims before the war in Iraq began? Better yet, how many Americans understood them four, five, six years into the war? The dynamics of empire compel us to defend and attack various people on the other side of the planet whom we scarcely understand or even know exist. So it was with Rome; so it is with us.

The narratives of imperial conquests are tragicomic Rube Goldberg sequences of the unexpected and excessive. For instance, in World War II we invaded Germany, for good reason, and discovered the concentration camps of the Holocaust. This led to overwhelming sympathy in America for the creation of a Jewish state in Palestine for humanitarian reasons. Our creation of Israel on a remote, unknown frontier in turn led to Jewish conflict with the Arabs, and radicalization of the Arab world. This radicalization eventually led to anti-American regimes in the Near East and, after some time, the American invasion and occupation of Iraq. Iraq brought us to the border of Iran, which we then accused of meddling in Iraq. Now some people want us to invade Iran, which in turn will push the frontier toward new enemies. Meanwhile, in the north, our newfound friends in Iraqi Kurdistan are fighting our NATO friends in Turkey over its oppression of the Kurdish populations on the Turkish side of the border. There is no end.

Another interesting sequence began with the 1953 CIA-sponsored overthrow of Iran's popular postwar prime minister, Dr. Mohammed Mossadegh, to create the autocratic monarchy of Shah Mohammad Reza Pahlavi, which we viewed as more suitable to our interests. In time, this led to the 1979 revolution in Iran and the creation of a radical Shiite Islamic state under the control of the Ayatollah Ruhollah Musavi Khomeini, which, in 1980, was attacked by his Sunni neighbor, Iraq's Saddam Hussein. The war, which was supported by loans from America's Sunni allies in Kuwait and Saudi Arabia, lasted eight years and bankrupted Iraq.

Hussein, wishing to solve his financial problems, invaded Kuwait in 1990. America rushed to the defense of Kuwait and Saudi Arabia in Operations Desert Shield and Desert Storm, stationing U.S. troops in Saudi Arabia, the land of the two most holy Muslim sites in Mecca and Medina. This inflamed the young Osama bin Laden, who, capitalizing on another American overseas adventure gone awry—the liberation of Afghanistan from the Red Army—built a terrorist organization in Taliban-led Afghanistan aimed at getting America out of Saudi Arabia. This in turn led to the 9/11 attacks against the World Trade Center in New York and the Pentagon, which in turn led to the American invasions of both Afghanistan and Iraq.

America subsequently withdrew its forces from Saudi Arabia and stationed them in occupied Iraq. Ironically, we have now negotiated a status of forces agreement (SOFA) with Iraq that requires us to abandon our bases there by the end of 2011. Meanwhile, back in Afghanistan, we are not only battling the Taliban but are also fighting elements within Pakistan supportive of the Taliban. Pakistan has become the new frontier.

As General McKiernan told Brian Williams of *NBC News* in June 2008, seven years into our occupation of Afghanistan:

> The insurgency is one of the great challenges. The issue with a very porous, uncontrolled, unsecure border area between Pakistan and Afghanistan is a challenge . . . I don't see when it ends. Our objective is to create a viable Afghan country with governance, with institutions that connect with the people, which are different requirements in different parts of Afghanistan. However, ultimately, strategically we have to look at this region. I think the answer to the question of what's—you know, what's the outcome that the world needs in this part of the world, this region, is bigger than Afghanistan.[17]

These sad chronologies illustrate not only the never-ending political entanglements and warfare necessary to maintain and expand a frontier, but also the unintended complications that bedevil such an enterprise. To the imperialist, the solution to the current problem is always the creation of a new one in the hopes that the new problem will be more manageable. Never is the imperialist content to chew what he has already bitten off. On the frontier of empire, the only perceived safety is in moving the frontier farther out. As it once was with the American Indians, so it is now with the world.

If we are ever to break our imperial habit, we must clearly define a difference between offensive wars, which expand the frontier, and defensive wars, which support allied governments overseas and our homeland. I personally would favor renaming the Department of Defense to a former appellation, the Department of War, so that Americans would no longer immediately associate all U.S. military activity with their defense. A change of terms is only a modest beginning but an important first step.

We must also develop a greater appreciation that the frontier line will always present us problems that cannot be solved, only managed. Keeping overseas problems *overseas* must become a central tenet of a nonimperial strategy. Never should we conflate America's homeland security with the interests of others across the oceans. Our friends and allies may have their lists of things they want to happen or not happen, but their interests need not be our interests. We would benefit from a policy of noninvolvement in most all the squabbles of the world. We can mediate solutions to problems without taking ownership or assuming responsibility for their resolution.

Finally, we must once again make military action the last resort. By this I do not just mean attempting diplomacy first, then going to war. Rather, a defensive strategy must recognize that the enemy will always strike the first blows in war. Letting the other side attack first and initiate the war was a principle of American military policy recognized from our foundation through the Cold War. Proactive war—as many now urge—is aggression, not defense.

A second requirement for avoiding the imperial trap is the reduction of the currently claimed powers of the executive branch in favor of the formula described in the Constitution. The Founding Fathers distilled the lessons of Rome into four simple rules: the Congress, the elected body closest to the people, will declare all wars; the armies for the war will be raised from among the people; the president will be the commander in chief of the armed forces; and the Senate will ratify whatever treaty ends the war, thereby establishing the new peace. This prescription for warfare, by spreading responsibility for war among the three nonjudicial entities, ensures that wars are not entered into lightly, and it institutionalizes the broadest support for any war declared.

Other constitutional passages empower Congress to "raise and support Armies," "maintain a Navy," "make Rules for the Government and Regulation of the land and naval Forces," and decide the necessary appropriations. Realizing the practical necessity of putting the armed forces of the nation under a single commander in chief, the president, the constitution's framers made clear that every other power over military policy was placed in the hands of the legislative branch. Especially, stated the Constitution's chief architect, James Madison, "The executive has no right, in any case, to decide the question, whether there is or is not cause for declaring war."

Through World War II this constitutional formula worked much as the framers desired. However, our post–World War II creation of an overseas empire meant maintaining a large permanent military establishment; consequently, greater freedom over military matters accrued to the executive branch under its command responsibilities. Worse still, the Cold War hair-trigger nuclear confrontation with the Soviets meant that exigent military measures might have to be taken before Congress could be assembled. Necessity required even more executive branch control over decisions of war and peace. Indeed, President Truman fought the Korean War without a declaration from Congress, and many in Congress felt that the Lyndon Johnson administration used the authorities given to him in the 1964 Gulf of Tonkin Resolution to fight an escalating and unpopular war in Vietnam that they never envisioned. After the repeal of the Gulf of Tonkin Resolution in

1971, President Richard Nixon claimed that his constitutional role as commander in chief of the armed forces gave him sufficient authority to continue to prosecute the Vietnam War as American soldiers were withdrawn.

To clarify the constitutional formula given the new Cold War conditions of permanent mobilization and persistent conflict, the Congress passed the War Powers Act of 1973 over Nixon's veto. Unfortunately, the resolution is inconsistent in what it says. Section 2 very clearly prohibits the president from committing acts of war without congressional preapproval, and allows the president to use the armed forces unilaterally only in response to enemy attacks on the United States:

> The constitutional powers of the President as Commander-in-Chief to introduce United States Armed Forces into hostilities, or into situations where imminent involvement in hostilities is clearly indicated by the circumstances, are exercised only pursuant to (1) a declaration of war, (2) specific statutory authorization, or (3) a national emergency created by attack upon the United States, its territories or possessions, or its armed forces.

Section 3 is far less explicit and can be construed as a loophole to allow the president to initiate acts of war by claiming that consultation was not "possible."

> The President in every possible instance shall consult with Congress before introducing United States Armed Forces into hostilities or into situations where imminent involvement in hostilities is clearly indicated by the circumstances, and after every such introduction shall consult regularly with the Congress until United States Armed Forces are no longer engaged in hostilities or have been removed from such situations.

Section 4 seems to provide the president even more latitude, providing him forty-eight hours of unrestricted use of American military forces if he deems, as suggested in Section 3, that consultation with Congress is not "possible."

> In the absence of a declaration of war, in any case in which United States Armed Forces are introduced—
>   (1) into hostilities or into situations where imminent involvement in hostilities is clearly indicated by the circumstances;
>   (2) into the territory, airspace or waters of a foreign nation, while equipped for combat, except for deployments which relate solely to supply, replacement, repair, or training of such forces; or
>   (3) in numbers which substantially enlarge United States Armed Forces equipped for combat already located in a foreign nation; the

president shall submit within 48 hours to the Speaker of the House of Representatives and to the President pro tempore of the Senate a report, in writing, setting forth—

(A) the circumstances necessitating the introduction of United States Armed Forces;

(B) the constitutional and legislative authority under which such introduction took place; and

(C) the estimated scope and duration of the hostilities or involvement.

Section 5 allows Congress to end wars that the president starts without authorization, though it is difficult to envision that this provision would, for practical and political reasons, be invoked:

> . . . at any time that United States Armed Forces are engaged in hostilities outside the territory of the United States, its possessions and territories without a declaration of war or specific statutory authorization, such forces shall be removed by the President if the Congress so directs by concurrent resolution.

Given the inconsistent language in the War Powers Act, there is little surprise that different people tend to interpret it in different ways. The executive branch, which resents the law's restrictions, tends to ignore the most restrictive passages in Section 2 and steers the discussion toward the consultation and notification provisions in the subsequent sections. Under the loosest interpretation, the president has the authority to initiate war and then dare Congress to renounce the war and the efforts and sacrifice of the American military. In this view, enunciated by the George W. Bush administration, the president is the "decider," and it is Congress' duty "to support the troops." (How the Obama administration will interpret the War Powers Act is an open question as this is being written. The executive branch, however, does not have extensive precedent for giving up powers claimed by its predecessors.)

Military officers swear an oath to "support and defend the Constitution of the United States against all enemies, foreign and domestic, [and] bear true faith and allegiance to the same." Under the loosest interpretation of the War Powers Act, they could construe orders from the president to initiate war with a foreign country without congressional authorization as being lawful orders that must be obeyed. Indeed, most officers I talk to believe that preemptive strikes ordered by the president are consistent with his constitutional authorities and do not require congressional approval or consultation.

This is an extremely dangerous situation for the American Republic. Once the president commits the nation to war by attacking another country, the attacked country has cause to fight us for as long as it chooses. For instance, were we to launch a preemptive strike against Iran's nuclear facilities—in other words, commit an act of war against Iran—that country would then have license to attack American targets as long as it maintained the will and means to do so. Ending the war would not be the president's choice, or even Congress' choice, but Iran's. Given the importance of peace in the Persian Gulf to the world's energy supplies, a limited strike against Iran could snowball into a requirement for a full-fledged invasion and occupation of Iran.

James Madison's admonition must be firmly enshrined in law. Unless the United States is under attack by a Russian nuclear first strike, our only current existential threat, there is simply no reason why the president cannot get congressional approval for any war or for any act of war that the administration may consider in the best interest of the nation.

Section 2 of the War Powers Act has it right. The duty of the president to respond to attacks on America or the American armed forces is absolute and unquestioned. Any other attacks, even against our friends and allies, insofar as they do not involve attacks on American forces, provide the president time for consultation with Congress to obtain the needed statutory authorizations and declarations to prosecute war. Never should the president initiate acts of war and risk placing the country in a state of war with another country without congressional authorization prior to the fact. While the Constitution does not prohibit the United States from initiating an offensive war, clearly congressional authorization should be mandatory before the commander in chief may direct the military to initiate acts of war.

Another subversion of the constitutional process, again magnified by a standing professional military, is the growing affinity of the military's career officer corps for the Republican Party. Despite admonitions from the senior leadership to carry forward the military tradition of staying politically neutral, most of the officer corps is now staunchly Republican and feels free to publicly state their preference for, if not allegiance to, that party. If there are Democrats among the officer corps, they hold their tongues in fear of being made outcasts.

This is new and dangerous. Officers of the World War II era were intensely nonpolitical, even to the point of never voting. Military obedience to civilian rule required unflinching service to all elected constitutional authorities regardless of party or policy. Cultural shifts that began in the Vietnam era changed the military ethos over time.

Prior to the 2008 election the Republican Party held the White House for twenty-eight of the previous forty years. In their executive capacity, Republican political appointees disproportionately shaped the modern professional military. Moreover, the Republican presidents positioned themselves as the proponents for a muscular armed forces and an aggressive military policies overseas. The armed forces became convinced that the Republicans were strong supporters of their institutional interests. The Democrats, on the other hand, were branded as the party of defeatism in Vietnam and military weakness in general. The Democrats became viewed as the party of the Congress, more concerned with the domestic concerns of their constituents than the projection of military power overseas.

One can only speculate about what this Republican bias portends. I cannot imagine an American Sulla marching into Washington to purge Congress of members who did not support his overseas imperial enterprises. More likely is that a Republican president will use the military's loyalty to, in effect, ignore and demote the Congress, ensuring that the requirements of empire are satisfied before the needs of the people—the imperial pattern of the caesars. Over time, it may become more difficult for Democratic administrations to find unbiased military advice as the professional military merely waits them out until a new Republican administration is elected.

The new "retired general as television pundit" phenomenon allows the professional officer corps an unprecedented access to shape American political opinion. How these hired guns will savage Democratic administrations is now unknowable, but clearly most of them will have Republican leanings. However the Republican-leaning officer corps manifests its political preferences, Americans must anticipate a new and uncomfortable era in civil-military relations.

Additionally worrisome are the legions of defense contractors and, lately, private military firms who swear no oath of allegiance to "support and defend the Constitution" but rather work for profit or their own private motives. President Eisenhower warned the nation of the perils of the military-industrial complex in his famous Farewell Address in 1961.

Our military organization today bears little relation to that known by any of my predecessors in peacetime, or indeed by the fighting men of World War II or Korea.

Until the latest of our world conflicts, the United States had no armaments industry. American makers of plowshares could, with time and as required, make swords as well. But now we can no longer risk emergency improvisation of national defense; we have been compelled to create a permanent armaments industry of vast proportions. Added to this, three and a half million men and women are directly engaged in the defense establishment. We annually spend on military security more than the net income of all United States corporations.

This conjunction of an immense military establishment and a large arms industry is new in the American experience. The total influence—economic, political, even spiritual—is felt in every city, every Statehouse, every office of the Federal government. We recognize the imperative need for this development. Yet we must not fail to comprehend its grave implications. Our toil, resources and livelihood are all involved; so is the very structure of our society.

In the councils of government, we must guard against the acquisition of unwarranted influence, whether sought or unsought, by the military-industrial complex. The potential for the disastrous rise of misplaced power exists and will persist.

We must never let the weight of this combination endanger our liberties or democratic processes. We should take nothing for granted. Only an alert and knowledgeable citizenry can compel the proper meshing of the huge industrial and military machinery of defense with our peaceful methods and goals, so that security and liberty may prosper together.[18]

In Eisenhower's day, private industry provided mainly military hardware. In the 1990s, however, under the Clinton administration, the notion developed that the Department of Defense could also "privatize" services that weren't essentially military in nature, such as housing and janitorial services, if "outsourcing" created cost savings for the government. Rather than use military personnel or civilian employees of the military to perform certain "nonmilitary" tasks, the military would let contracts to private firms to provide the needed services.

The Bush administration expanded the contracting of "services" by several orders of magnitude. The then chairman of the House Defense Appropriations Subcommittee, John Murtha (D-PA), estimated that the United States employed about 126,000 contractors in Iraq, with "some of them making more than the secretary of defense."[19] Representative Jan Schakowsky (D-IL), of the House Intelligence Committee, estimated that war contractors represented 40 percent of the cost of our occupation in Iraq.[20] Most of these contract workers prepare meals in mess halls, clean latrines, or drive trucks—common labor.

Many of my mostly army-officer students who have served in Iraq openly question whether the vested interest that these contractors have in seeing their income stream continue works against America's interest. As one officer said, "These guys would be happy if the war went on forever." The March 2008 awarding of ten-year, $150 billion service contracts to a consortium of defense firms to house, feed, and otherwise support the basing of U.S. forces in Iraq means that for all practical purposes the defense contractors want to see their profitable businesses continue indefinitely.

However, not all of the contractors serving in Iraq and Afghanistan perform essentially nonmilitary support tasks such as KP duty. It's estimated that twenty thousand to thirty thousand of these service workers are armed security guards, generally former U.S. combat arms and special forces soldiers, performing inherently military tasks once done by uniformed military policemen or other armed government officers. At least $4 billion in taxpayer money has been spent on these armed security contracts.[21]

Blackwater and two other private security firms, DynCorp and Triple Canopy, form the backbone of America's new private armies. Blackwater, which was founded in 1998 to support military training and had federal contracts valued at less than $1 million in 2001, has received over $1 billion in federal contracts from 2002 onward, mostly to provide security services for the State Department.[22] These contractors have been implicated in unwarranted killing of Iraqis, torturing of prisoners, and illegal "renditioning" of terrorist suspects to countries where they will be tortured. As contractors, they swear no oath "to support and defend the constitution" and are not accountable for their actions under the Uniform Code of Military Justice, the legal codes that pertain to the members of the U.S. armed forces.

Some have suggested that these firms could form the nucleus of a new Praetorian Guard, the extralegal palace guard that Octavian used to protect his rule in Rome. Though that is an exaggeration, it is worrisome that the American taxpayers are funding the creation of a sizable paramilitary force that resides outside the judicial and administrative constraints that we have wisely imposed on our military and police forces.

Worse, all these contracts for overseas contractors are essentially executive branch fiats not subject to the oversight and congressional interest that attends government contracts awarded for stateside work. Bush administration political appointees fired and overruled senior nonpartisan civilians who, in their capacities as Department of the Army and Department of Defense auditors, questioned billion dollar payments to Kellogg, Brown, and Root, which is a subsidiary of Halliburton, the Texas oil services company formerly run by Vice President Dick Cheney.[23]

Perhaps the most damning assessment of the Bush administration's profligate and unsupervised expenditures to contractors comes from a June 16, 2008, letter from Congressman Henry Waxman (D-CA), chairman of the House Committee on Oversight and Government Reform, to Claude M. Kicklighter, the inspector general of the U.S. Department of Defense:

> I am writing to seek your assistance in investigating potentially thousands of criminal cases involving fraudulent contracts in Iraq . . .
>
> On May 22, 2008, your deputy, Mary Ugone, testified before the Oversight Committee and released a report assessing the Defense Department's oversight of billions of dollars in procurement expenditures in Iraq. The primary finding in the May 22 report was that the Defense Department "did not maintain adequate internal controls over

commercial payments to ensure that they were properly supported." The report estimated that $7.8 billion out of a pool of $8.2 billion in commercial payments "did not meet all statutory or regulatory requirements."

Of this amount, the report estimated that $1.4 billion in commercial payments "lacked the minimum documentation for a valid payment, such as properly prepared receiving reports, invoices, and certified vouchers" and thus "do not provide the necessary assurance that funds were used as intended."

During the course of the Iraq War . . . the Defense Department has repeatedly failed to take reports of waste, fraud, and abuse seriously. This was demonstrated once again when the Department failed to acknowledge the critical problems highlighted by your report and refused to send Pentagon officials to testify voluntarily at the Oversight Committee's May 22 hearing. It is difficult to have confidence that the Department will take adequate steps to address the full magnitude of the potential criminal fraud.

We may never know to what extent the Bush administration awarded contracts or provided financial profits to its political allies that were inconsistent with the nonpartisan interest of the American people. Every war has created contracting scandals. But Iraq is clearly the worst. Operating without fear of meaningful third-party review, the Bush administration transferred great sums of money to contractors with little assurance that the money was properly spent.

The third requirement for the defense of the American Republic is to dramatically reduce the defense budget, the share of the gross domestic product spent by the Department of Defense and other related agencies, and increase spending for domestic needs. The health of the republic begins at home, not abroad. As President Eisenhower stated in 1953:

Every gun that is made, every warship launched, every rocket fired signifies, in the final sense, a theft from those who hunger and are not fed, those who are cold and are not clothed. This world in arms is not spending money alone. It is spending the sweat of its laborers, the genius of its scientists, the hopes of its children. The cost of one modern heavy bomber is this: a modern brick school in more than 30 cities. It is two electric power plants, each serving a town of 60,000 population. It is two fine, fully equipped hospitals. It is some fifty miles of concrete pavement. We pay for a single fighter plane with a half million bushels of wheat. We pay for a single destroyer with new homes that could have housed more than 8,000 people . . .

This is not a way of life at all, in any true sense. Under the cloud of threatening war, it is humanity hanging from a cross of iron.[. . .] Is there no other way the world may live?[24]

Eisenhower, of course, was a war hero, not a pacifist. Forced by the challenge of the Soviet Union to emplace a global defense, Eisenhower spent what needed to be spent. During Eisenhower's eight years in office, which coincided with the height of the Cold War, America spent slightly more than 10 percent of its gross domestic product on defense.[25]

Over time, the Soviet threat diminished and the American economy grew. Defense spending fell to 5.8 percent of the gross domestic product during the Reagan decade of 1980 to 1989.[26] The end of the Cold War further reduced defense needs, and during the Clinton years defense spending dropped to 3 percent of the gross domestic product in fiscal years 2000 and 2001, in the neighborhood of $300 billion per year.

To fight the global war on terror, the Bush administration, with congressional approval, increased "baseline" defense spending to 3.8 percent of the gross domestic product. The $481 billion baseline defense budget for 2008, itself an increase of 62 percent over the 2001 figure, is augmented by periodic, off-budget "emergency" supplemental appropriations for operations in Afghanistan and Iraq that the Congressional Budget Office estimates total $752 billion from 2001 through February 2008. Additional approved supplemental requests for 2009 lifted the total to close to $900 billion for these two wars through the end of the Bush administration, with much more programmed to be spent during the Obama presidency. The Obama administration pledges to reshape the budget process to include the cost of our current wars in the fiscal year budget, no longer "hiding" them as off-cycle supplemental appropriations.

The decision to respond to 9/11 with an offensive war strategy has cost the nation nearly $2 trillion as of early 2009, with much more to come. The cost of the war is not fully captured by the $900 billion in wartime supplementals. Also needed for the war effort, baseline defense spending—including nuclear weapons programs in the Department of Energy—rose from $316 billion in fiscal year 2001, which ended in September 2001, to $535 billion in fiscal year 2009, for another near trillion dollars in military spending.

These added offensive force structure costs—over $200 billion per year above the cost of our defensive force structure cost in 2001—will continue as long as America desires to maintain an offensive war capability, whether or not we actually use it. (The cost of a post-9/11 defensive military strategy, the path not taken, can perhaps be gauged by the $17 billion in supplemental defense spending that paid for global war on terror operations in 2001, including the U.S. capture of bin Laden's base at Tora Bora.[27])

Remarkably, none of this increased defense spending has been offset by increased taxes. Unprecedented in the history of the republic, the president refused to raise taxes for the wars he urged and commanded. Instead, the Bush administration cut taxes and financed the increased defense expenditures solely through borrowing. A $236 billion budget surplus in fiscal year (FY) 2000 turned into a $413 billion budget deficit in 2004. Deficits due to the defense buildup and costs of the wars persist.

Even before the mortgage meltdown and the recession ballooned the 2008 deficit to over a trillion dollars, the Bush White House had predicted the 2008 budget deficit at $410 billion even with a "normal" economy. The chronic global war on terror deficits have caused the national statutory debt limit of $5.95 trillion in 2002 to be raised to $9.815 trillion in September 2007. Consequently, the on-budget net interest expense of $206 billion in 1991 is estimated to rise to $382 billion in fiscal year 2009.[28] This figure amounts to about 2.5 percent of the U.S. gross domestic product, or over 12 percent of the $3.1 trillion federal budget. Fighting the global war on terror offensive strategy on credit has saddled future administrations, not to mention generations, with significant debt burdens.

Unlike the debt that America incurred for World War II, which was wholly domestic debt borrowed from Americans through instruments such as war bonds, the current U.S. deficit is increasingly funded overseas by parties that often have interests much different than ours.

This is dangerous for two reasons. First of all, fighting wars on credit raised overseas makes it impossible for the American people to accurately judge whether a war is worth the cost. Increased taxes and domestic borrowing for war both restrict domestic consumption in ways that drive home to the American people the true economic cost of the war. One cannot do cost-benefit analyses without gauging "cost." Bush's telling the American people after 9/11 to "go shopping" as their contribution to the war effort contrasts weakly, both morally and economically, with Roosevelt's pleas that Americans "buy bonds" during World War II. Financing war through foreign borrowing does not adequately communicate "cost" to decision makers and the electorate at large. It is doubtful that Americans would have approved of the tremendous expense of an offensive, military-centric global war on terror had the bill been paid as it was incurred.

The second danger is that much of this American debt is owned by the Chinese, our principal rival going forward in the twenty-first century, and Saudi Arabia, the source of much of the radical Islamic beliefs and financial support that animate and enable our foes in the global war on terror. America's policy of military expansion is being bankrolled by regimes whose interests often conflict with ours. So far, neither of these creditors has overtly used its willingness to buy American debt as a means to control American policy, but one cannot rule out a demand for a quid pro quo in the future.

It is fair to say that our current military expansion will last only as long as the Chinese and Saudi Arabians wish to pay for it, not a day longer. When we Americans must again shoulder our own military expenditures, we will likely be far less ambitious in our aims and more cost-conscious in our means.

It would be wrong to conclude that because our current military expenditures are, as a percentage of the gross domestic product, lower than they were in the 1950s they are therefore affordable. For the past forty years, federal spending has remained fairly constant, at around 20 percent of the gross domestic product.[29] Higher rates of spending cause a conservative revolt against rising taxation; lower levels of spending create popular pressure for more public services. The real difference between the 1950s and the current situation is the rapidly growing elderly and retiree population and their income and medical needs.

During the 1950s the nation's demographics allowed for very modest federal expenditures for retirees. Eisenhower could spend over half the federal budget on defense and still have ample money to fund a vast federal highway program, the building of universities and colleges, scientific research, and a variety of federal programs that irrefutably strengthened America domestically. Moreover, federal expenditures during Eisenhower's administration averaged only about 18 percent of the gross domestic product, allowing room for states and localities to raise taxes for their own programs.

Now, with an aging American population, the federal government lives in a budgetary straitjacket. Statutory Social Security, Medicare, Medicaid, and veterans benefits now consume over half the federal budget and more than 10 percent of the gross domestic product, and these percentages are virtually certain to continue to increase over the next several decades. Mandatory interest payments are another 2.5 percent, "baseline" military expenses another 4 percent, and global war on terror "supplemental" expenses another 1 percent. Only about 3.5 percent of the gross domestic product supports all the other departments and functions of the national government, including agriculture, education, transportation, energy, and assistance to states. Consequently, our national infrastructure is crumbling, our educational standards are falling to second-tier status among industrialized nations, and other discretionary budget items are reduced or eliminated.

Making budgeting contraints even worse for the military, in order to fund the current increases in Social Security, Medicare, Medicaid, veterans, and interest expenditures, the Congressional Budget Office baseline mandates that nondefense discretionary expenditures fall to 3 percent of the gross domestic product by 2015 while the defense share of the gross domestic product falls to 3.1 percent.

Given the budget realities, the current military offensive is a short-lived policy that we cannot afford to continue. Americans are not going to throw our aged parents onto the streets or deprive them of needed medical care in order to raise funds for elective wars on the other side of the planet.

Sooner than later, the fiscal irresponsibility of the current expansionist policy will end, perhaps because our creditors will demand higher interest rates or quit lending to us altogether. Perhaps successor administrations will bring spending priorities more in line with domestic needs. Perhaps a prolonged recession will make foreign wars unaffordable. Whatever the mechanisms for change, the enduring truth is that the military costs of empire are unaffordable and inconsistent with democratic governance. No nation can long endure the cost of governing others. Rome couldn't, the European colonialists couldn't, the Soviets couldn't, and neither can we.

Expending vast sums on unnecessary military adventures robs a democracy of the wealth it needs to ensure domestic prosperity and international economic competitiveness. Wasting a few percentage points of gross domestic product on war rather than investing them productively in the domestic economy, given the compounding effect of time, makes a vast and decisive difference in the prosperity and well-being of the people and the nation. England is better off for having shed its empire; Russia is poorer for having attempted to keep theirs. How will America decide?

Chapter 15

Offensive War, Governance, and Empire

In the preceding chapters, I have attempted to weave a thread among several interrelated schools of thought that unfortunately tend to be viewed as separate fields of activity.

American military history produces an invaluable record of successful offensive wars, which, when examined in their entirety, yield excellent baseline data that provide sobering estimates of the costs and benefits of any proposal to emplace new governance over a foreign population. No military on Earth has waged offensive war as effectively as has the American military over the past four centuries. The army's historical experience provides critical lessons concerning the need for attrition and casualties, the imperatives of counterinsurgency campaigns, and the organization and employment of military governance units. Yet we discount our history of success as being irrelevant to current warfare.

Because human nature is slow to evolve, more has remained constant in warfare than has changed during the time frame from the Indian Wars through the wars in Iraq and Afghanistan. Men organized in societies still fight wars, and weapons confer advantages, but the processes of war are largely human and not technologic. The lessons of the past are still valid and instructive today.

Consequently, the military must revise its doctrine so that it clearly summarizes the enduring truths reflected in the empirical record. The historicism of Clausewitz has served us poorly in the thirty years since military theoreticians made him the centerpiece of our new joint doctrine. Center of gravity analysis confounds military planners and leads only to the wishful thinking, implicit in our doctrine, that a general's strategies and decisive battles can reduce or avoid the need for attrition warfare. There is no historical proof of such a claim.

Similarly, the military must realize that offense, defense, and stability are not just types of operations but indeed types of wars. Offensive wars are fundamentally different from defensive wars in both their aim and conduct. Imposing new governance on a resistant population is the necessary aim of offensive war toward which all operations must be directed. Winning the force-on-force battle only enables the attacker access to the population and must be circumscribed within that context. Defensive wars are fought to preserve established governance either through deterrence or direct defensive force-on-force combat and do not require establishing new forms of governance anywhere.

The distinction between the terms "occupation" and "liberation" is not in the mind of the speechwriter trying to put the right political spin on a military operation. Rather the distinction is a product of the initial conditions of the war and the war aims of the invading country, and must be reflected in the civil-military force structure and operations of the invading force. We knew this in 1940 when the army published Basic Field Manual 27-5, "Military Government," for use in World War II. We had no such doctrinal understanding when we invaded Iraq and Afghanistan, and predictably the result was chaos and insurrection.

The military is responsible for knowing its history, culling its lessons, and promulgating sound military doctrine. Our failure to do so in the post-Vietnam era is an indictment of the military as a profession, which has a duty to maintain its appointed body of knowledge in the interest of its client, the American people and government.

There is no sugar-coating it. The military entered the current wars in Iraq and Afghanistan in willful negligence of both the kinds and scopes of operations that would be required to win them. Whatever portion of that negligence can be attributed to administration pressures is debatable, but the evidence is clear that the military entered the new millennium without any meaningful offensive war doctrine and did not develop one subsequent to 9/11, even though the Bush administration was clear in its intentions to go on the offensive. Prudence would demand that the institutional failures that created this state of affairs be fully investigated and corrected. However, with so many sins to hide and reputations to protect, there is little evidence that the army and the rest of the joint military community are yet up to that task.[1]

Wars, however, are not discrete events unlinked to the larger and longer-term military policies of the nation. A military that is structured and trained for a policy of strategic defense will likely not be able to transform itself for offensive war without time-consuming changes to its doctrine, force structure, and training. The very term "Department of Defense" causes confusion, implying to people within the military and outside of it that our actions are primarily intended for the purpose of defending the United States. Perhaps during the Cold War this was true, but at the turn of the new millennium the purpose of U.S. military action had become doctrinally offensive: invading foreign countries, deposing their governments, and then establishing new governance in accordance with our national interests.

The recent neoconservative military expansion into the Middle East and central Asia ultimately does not protect America; it attempts to spread Americanism, and at great cost, to areas of the world that are, for a variety of reasons, resistant to it. Not only have our wars in Iraq and, to a lesser degree, Afghanistan been unpopular in the region, many of our allies in the Cold War collective defense have either refused to participate in these wars or have committed only small numbers of troops and support.

As America shifts from a Cold War defense to a global-war-on-terror offense, we are viewed less favorably worldwide. The city on the hill loses its luster as it compels, rather than attracts, others to its domain. Former friends cast a wary eye; potential enemies arm in fear that they may be next.

And the constant military mobilization made necessary by these elective overseas wars and occupations changes, for the worse, the fabric of the United States. From the first days of the individual thirteen colonies, the American armed forces have always existed within a democratic context that today is unraveling.

The permanent military establishment made necessary by the Cold War challenges the constitutional framework of the nation more surely than does any current foreign enemy. The end of conscription and the professionalization of the U.S. armed forces further isolate the military community from the citizenry it serves. Consequently, we see the all-volunteer military establishing a political identity with the executive branch of government, and especially the Republican Party, not reflective of the population at large. The checks and balances of the constitutional formula and the rule of law are being replaced by increasing executive branch prerogative, disdainful of congressional oversight and interference. The rule of law, once a cornerstone of the American political system and even the conduct of war, gives way to naked power and rule by force.

The financial costs of our far-flung military offensives are borrowed, and the debts are foisted off on future generations in ways that would be unimaginable to the citizens who built the wealth of this nation. The quest for empire, the compulsion to keep extending the distant frontier outward at all costs, brought down the Roman Republic. Empire could be the undoing of our republic as well.

It is all connected. Governance and military policy are inescapably linked, both overseas and at home. Just as the military needs to refocus its understanding of warfare to consider governance first and foremost, so too must the federal government and the citizenry carefully weigh the impact of American military policy on the health of the republic. It is difficult to argue that our recent military offensive strategy has actually strengthened America domestically or overseas. Only the deliberate exaggeration of the enemy, the creation through propaganda and fear of an existential threat that does not in actuality exist, underpins our expansionist folly.

There is a general feeling at home and abroad that America over the past several years has set a wrong course that must be righted. The sheer incompetence of our efforts in Afghanistan and Iraq, for which the senior military leadership must share blame with the then-incumbent administration, is unworthy of a country that so lavishly supports its armed forces and fills the role of the world's only superpower, functioning as Thomas Hobbes' and Thomas Barnett's leviathan.

But the incompetence is born of hubris, and the hubris born of pretension, both of which betray the humble origins of the American experiment. This hubris insults the democratic aspirations of those worldwide who believe that freedom is the only antidote to naked power and tyranny. This betrayal of America's finest ideals is the true failing of the current military policy. The notion that the city on the hill would embark on a policy of conquest is enormously unsettling to Americans who aspire only to perfect the city and foreigners who looked to America for an example of how they might better themselves.

It is not too late for America to change course. A policy of strategic defense, America's international policy for nearly all of our history, has served the dual aims of making America prosperous and strong at home and spreading democracy abroad. Against the background of progress that America has brought the world, perhaps future generations will remember America's good intentions in Iraq and Afghanistan more than our half-truths, missteps, and imperious behavior. As a gesture of our new sincerity, we must allow these two countries true sovereignty, let them choose their own course, and offer our friendship whatever their choice. As one general officer recently described our path forward, "We need to support what is acceptable to them and tolerable to us, not the other way around."[2]

Forswearing future offensive wars, unless authorized by a recognized international body, and pledging to abide by the rule of law would enhance America's image and "soft" power overseas, as well as make American military power more acceptable internationally when it is truly needed to respond to aggression.

A militarily defensive strategy would also save the American treasury the enormous burdens of maintaining and operating a large and expensive offensive military establishment and thereby free up resources for the restoration of the American economy and infrastructure, as well as a wide range of investments in education and health care. Over time, the republic would heal and strengthen. All human systems make errors of judgment. The real tragedy is to persist in error rather than acknowledge it and make the necessary corrections.

This treatise began with an inquiry into the nature of offensive war and concludes with a study of empire. It began by analyzing military necessity and ends by examining the nature of the republic. All of these discussions are inexorably linked, and we must begin to acknowledge the connections, as did our Founding Fathers.

Military competence in the service of empire is not a virtue; it is a crime and a tragedy. On the other hand, military service in the name of the republic, as described by the Constitution, must be affirmed as the righteous devotion and aspiration of all who wear the uniform. Similarly, the civilian officials, either elected or appointed, who govern military affairs must put the needs of the republic, as an enduring institution, above the transient desires and passions of the day. The preservation of the republic, its ideals and proper functioning, must rise above any other agendas, foreign or domestic.

Here we have collectively failed. We must mend our ways, lest we bear the judgment of history.

FM 27-5, "Basic Field Manual: Military Government" (1940)

I include FM 27-5 because it is the clearest and most concise statement available of the policies and procedures of military governance. Field Manual 27-5 was first published in 1940 and was the army's first approved doctrinal manual on the subject.

Despite the lack of official published doctrine prior to 1940, the army had studied and taught military governance procedures for the century prior to World War II. In the early twentieth century, for instance, the Command and General Staff College at Fort Leavenworth taught classes on the subject, based on our Progressive Era experience, and even had students plan a hypothetical occupation of Mexico in a classroom training exercise. Field Manual 27-5, 1940, is best viewed as the first time the army officially codified its knowledge on the subject.

The field manual is historic because it necessitated, in its opening section, the creation of the force structure and training base necessary for our postwar occupations in Europe and Japan.

I selected the 1940 edition—rather than the December 1943 edition, which was a joint army and navy publication, or the 1947 version—simply because the original is shorter, more pointed in its discussion, and more easily understood by a nonmilitary reader. Only twenty-three pages, the main document is a pithy and quick read, especially compared to the doctrinal tomes published by today's military. (I have omitted the appendix, which contains lengthy, often repetitive detail directed toward the legal profession.)

A simple reissuing of the 1940 version of FM 27-5 in, say, December 2001 could have avoided many of the early, perhaps irrecoverable, failings of our occupations in Iraq and Afghanistan. Page 1 makes the commanding general of the theater of operations—not some multinational or interagency organization, or the State Department— responsible for occupation governance. Page 2 directs the army to raise and train the force structure, that is, soldiers and units, necessary for occupation governance. In Iraq and Afghanistan, the army thought that someone else would send the people and assume the mission. No one yet has. Page 5 requires separate personnel for military government and recognizes that combat units are not suited for the task, a direct contradiction of the army's current Full-Spectrum Operations doctrine. Pages 7 through 9 enumerate the various responsibilities of military government, everything from public works and utilities to public safety and economic reconstruction. The army applauded Maj. Gen. Peter Chiarelli's "discovery" of these categories of civil affairs operations when he published his findings in 2005, the third year of the war in Iraq.[1] Pages 10 through 12 discuss the organization of military governance and how army military governance units map to preexisting political subdivisions. The pages find their modern incarnation in the provincial reconstruction team approach so belatedly developed in Iraq and Afghanistan.

The largest section of the field manual, pages 12 through 19 and the thirty-five-page appendix, detail the procedures and ordinances that the military government will institute to establish a rule of law over the occupied nation. Rule of law, a concept central to the army's successful occupation policy from the Mexican War through World War II, was sadly lacking in Iraq after the American invasion in 2003 and remained a central issue of contention during the 2008 status of forces negotiations.

The final four pages of the field manual outline the commanding general's initial proclamation of occupation, informing the public of its purpose and policies, and the initial ordinances of the occupation; both are to be published in common language for the affected population at the commencement of military governance. These proclamations form a sort of contract between the military government and the occupied nation. No such proclamations and ordinances were provided to the Iraqi people in March 2003 and chaos resulted. Also, there remains great confusion in Iraq, America, and the world concerning our true reasons and intentions.

FM 27–5

BASIC FIELD MANUAL

❧

MILITARY GOVERNMENT

Prepared under direction of
The Judge Advocate General

UNITED STATES
GOVERNMENT PRINTING OFFICE
WASHINGTON : 1940

WAR DEPARTMENT,
Washington, *July 30, 1940.*

FM 27–5, Military Government, is published for the information and guidance of all concerned.

[A. G. 062.11 (5–23–40).]

By order of the Secretary of War:

G. C. MARSHALL,
Chief of Staff.

Official:

E. S. ADAMS,
Major General,
The Adjutant General.

TABLE OF CONTENTS

TABLE OF CONTENTS

BASIC FIELD MANUAL

MILITARY GOVERNMENT

SECTION I

GENERAL

■ 1. SCOPE.—This manual deals primarily with the policy of military government and its administration. Chapter 10, FM 27-10 (now published as BFM, Vol. VII, pt. two), deals primarily with the legality of military government. The chapter cited tells what may legally be done, this manual what it is advisable to do.

■ 2. PURPOSE.—The purpose of this manual is to furnish a guide for the War Department in planning military government, for commanders and their staffs in its establishment, and for personnel of all ranks in its operation.

■ 3. DEFINITION.—Military government is that form of government which is established and maintained by a belligerent by force of arms over occupied territory of the enemy and over the inhabitants thereof. In this definition the term *territory of the enemy* includes not only the territory of an enemy nation but also domestic territory recovered by military occupation from rebels treated as belligerents.

■ 4. OCCASION.—The military occupation of enemy territory suspends the operation of the enemy's civil government therein. It therefore becomes necessary for the occupying power to exercise the functions of civil government in the maintenance of public order. Military government is the organization through which it does so.

■ 5. AUTHORITY.—The exercise of military government is a command responsibility, and full legislative, executive, and judicial authority is vested in the commanding general of the theater of operations. By virtue of his position he is the military governor of the occupied territory and his supreme authority is limited only by the laws and customs of war.

particularly those set out in chapter 10, FM 27–10, and by such instructions as he may receive from higher authority.

■ 6. PLANNING.—The Personnel Division (G–1) of the War Department General Staff is responsible for the preparation of plans for and the determination of policies with respect to military government. The personnel section (G–1) of the staff of the commanding general, theater of operations, will, in advance of the necessity for the establishment of military government, make such further and more detailed plans therefor as may be necessary.

■ 7. PROCUREMENT OF PERSONNEL.—Pursuant to the plans made as provided in paragraph 6, the necessary personnel, commissioned, warrant, and enlisted, will be selected and procured. If the war is such as to require the maximum man power of the United States for combat and if hostilities are still in progress or only temporarily suspended, personnel for military government should be selected from those unsuited for combat duty by age, physical disability, existence of dependents, or otherwise. They should be selected by reason of knowledge or experience particularly fitting them for the work to which it is anticipated that they will be assigned. Those having experience in a former military government, in our own Federal government, or in that of a state, county, or city, or in public utilities, or as lawyers, physicians, civil engineers, and those well acquainted with the country to be occupied by former residence or travel therein are especially valuable. Knowledge of the language of the inhabitants is highly desirable but not indispensable. Civilian citizens of the United States should not be employed in military government. If particularly suitable individuals, not in the military service, are available and desired, they should be commissioned, warranted, or enlisted. Insofar as tactical requirements permit, selection and procurement of personnel will be made long enough in advance of the necessity for their use to permit their adequate instruction and training.

■ 8. TRAINING.—The Personnel Division (G–1) of the War Department General Staff plans and supervises the instruction and training of the personnel necessary for military government. In accordance with such plans and subject to such supervision, the personnel section (G–1) of the staff of the commanding general, theater of operations, makes such further and more detailed plans as may be necessary with respect to such instruction and training, so far as they may be carried on in that theater, and supervises them. So far as time and available facilities permit, the instruction will cover the law and practice of military government, the history of such governments in the past, and the language, geography, history, economics, government, and politics of the country to be occupied. In advance of the need for its use, the Military Intelligence Division (G–2) of the War Department General Staff will furnish data on the subjects last mentioned which may be used for instructional purposes. These data will be furnished to the theater commander and distributed to all officers and warrant officers employed in military government and to such enlisted men as may need them.

SECTION II

POLICIES

■ 9. BASIC.—Any plan of military government should conform to the following basic policies:

a. Military necessity.—The first consideration at all times is the prosecution of the war to a successful termination. So long as hostilities continue, the question must be asked, with reference to every intended act of the military government, whether it will forward that object or hinder its accomplishment. The administration of military government is subordinate to military necessities involving operations, security, supply, transportation, and housing of our troops. If hostilities are suspended by an armistice or otherwise, all plans and dispositions must be made so that the troops may resume hostilities with the least inconvenience to themselves and to the operations of the military government, and, above all, under conditions most conducive to a successful termination of the war.

b. Welfare of the governed.—Subject only to the foregoing, military government should be just, humane, and as mild as

3

practicable, and the welfare of the people governed should always be the aim of every person engaged therein. As military government is executed by force, it is incumbent upon those who administer it to be strictly guided by the principles of justice, honor, and humanity—virtues adorning a soldier even more than other men for the very reason that he possesses the power of his arms against the unarmed. Not only religion and the honor of the Army of the United States require this course but also policy. The object of the United States in waging any war is to obtain a favorable and enduring peace. A military occupation marked by harshness, injustice, or oppression leaves lasting resentment against the occupying power in the hearts of the people of the occupied territory and sows the seeds of future war by them against the occupying power when circumstances shall make that possible; whereas just, considerate, and mild treatment of the governed by the occupying army will convert enemies into friends.

c. *Flexibility.*—A plan for military government must be flexible. It must suit the people, the country, the time, and the strategical and tactical situation to which it is applied. It must not be drawn up too long in advance or in too much detail, and must be capable of change without undue inconvenience, if and when experience shall show change to be advisable.

d. *Economy of effort.*—Every man engaged in military government is withdrawn either from the combatant forces or from productive labor at home. All plans and practices should be adopted with a view of reducing to the minimum consistent with the proper functioning of military government the number of the personnel of our Army employed in that government and the amount of work required of them.

e. *Permanence.*—The system of military government should be planned so as to provide permanence for the duration of the occupation, and thus insure continuity of policy. Frequent changes in personnel or policy are to be avoided.

■ 10. SECONDARY.—As corollaries of the foregoing basic policies, the following policies should be followed in the planning and operation of military government. Experience in former military governments shows the desirability of so doing.

4

a. Supremacy of the commanding general.—It follows from the basic policy of military necessity (par. 9*a*) that the commanding general of the theater of operations, who is responsible for the success of the army there operating, must also have full control of military government therein. (See par. 5.)

b. Separate personnel for military government.—It also follows from the basic policy of military necessity (par. 9*a*) that, so long as hostilities continue, personnel of combatant units should not be charged with any responsibility for military government, and separate personnel should be provided therefor. If hostilities are suspended by an armistice or otherwise, duties with respect to military government may properly be imposed upon the personnel of combatant units only if the probability of a resumption of hostilities is extremely remote.

c. Retention of existing civil personnel.—It follows from the basic policy of economy of effort (par. 9*d*) that, so far as reliance may be placed upon them to do their work loyally and efficiently, subject to the direction and supervision of the military government, the executive and judicial officers and employees of the occupied country, its states, provinces, counties, and municipalities should be retained in their respective offices and employments, and held responsible for the proper discharge of their duties. If unwilling to continue in the performance of their duties, they may, as a matter of international and military law, be compelled to do so, any law of their own country to the contrary notwithstanding, provided the services required do not involve them in military operations against their own country. Whether officers and employees shall be required to serve against their will is a question of policy only, to be decided by the proper authorities of the military government in the light of the circumstances. The personnel of the military government should, so far as possible, deal with the inhabitants through the officers and employees of their own government.

d. Avoidance of changes in existing laws, customs, and institutions.—The existing laws, customs, and institutions of the occupied country have been created by its people, and are presumably those best suited to them. They and the officers and employees of their government are familiar with them, and any changes will impose additional burdens upon the military government. Therefore, it follows from the basic policies of welfare of the governed (par. 9*d*) and economy of effort (par. 9*d*) that the national and state laws and local ordinances of the occupied territory should be continued in force, the habits and customs of the people respected, and their governmental institutions continued in operation, except insofar as military necessity (par. 9*a*) or other cogent reasons may require a different course.

   e. Retention of existing political divisions.—The people of the country occupied, and the officers and employees of their government are familiar with the existing division of the country into states or provinces, counties or departments, and cities or communes. The laws and ordinances in force in one of these divisions are often different from those in another and would be unsuitable in that other. Disregard of these divisions will be disadvantageous to the people and place additional burdens upon the military government. It therefore follows from the basic policies of welfare of the governed (par. 9*b*) and economy of effort (par. 9*d*) that the military government should be so organized that its territorial divisions coincide with those previously existing, except insofar as military necessity (par. 9*a*) or other cogent reasons require a different course.

SECTION III

CIVIL AFFAIRS SECTION OF THE STAFF OF THE COMMANDING GENERAL, THEATER OF OPERATIONS

■ **11. ESTABLISHMENT.**—Whenever the establishment of military government is contemplated, the commanding general of the theater of operations will create a section of his staff, called the civil affairs section or the office for civil affairs, or by a similar title. Thereafter, pursuant to the policy of separate personnel for military government (par. 10*b*), though they will consult together whenever necessary and coordinate their policies, the general and special sections of the military staff of the commanding general will have no responsibilities with respect to civil government, and the civil affairs section will have no responsibilities with respect to military matters.

6

■ 12. Officer in Charge of Civil Affairs.—*a. Position, qualifications, and rank.*—The officer in charge of civil affairs will be the head of the civil affairs section of the staff of the commanding general of the theater of operations. He will be most carefully selected for his knowledge of, experience in, and qualifications for military government, and should have rank commensurate with his important responsibilities.

b. Duties.—The duties of the officer in charge of civil affairs are—

(1) Adviser to the commanding general and the staff on matters pertaining to the administration of civil affairs in occupied territory.

(2) Handling for the commanding general, in accordance with his approved policies, matters in connection with such military supervision or control of civil affairs as is necessary.

(3) Supervision of civil affairs in subordinate territorial commands.

■ 13. Organization.—The civil affairs section will be organized in such manner as the officer in charge of civil affairs, subject to the approval of the commanding general, shall decide. A type organization of such a section by departments follows, suitable for the occupation of a territory of considerable size for some time; but departments may be combined or omitted, or the work of the section otherwise divided, as the exigencies of the particular occupation may require or as experience may show to be advisable. If the country occupied has a well organized civil government whose personnel remain at their posts and perform their duties satisfactorily, the volume of work falling upon the military government will be much diminished and it may be possible to combine some of the departments or omit them.

a. Public works and utilities.—This department will supervise railroads, canals, harbors, rivers, lighthouses, buoys, roads, bridges, busses, trucks, street railways, gas, electricity, water works, sewerage, drainage, irrigation, forests, and the like. The officer in charge of it should be an engineer by profession.

b. Fiscal.—This department will supervise the financial affairs of the occupied territory, including taxes, customs, disbursements, coinage, currency, foreign exchange, banks, stock exchanges, and similar matters. It will receive any taxes, contributions, fines, or penalties collected by the military government. It will establish an adequate audit of the financial transactions of the military government and such additional audit of the financial transactions of the civil government as may be necessary for the purposes of military government. The officer in charge of this department should have had wide experience in financial matters.

c. Public health.—This department will exercise supervision over the public health, including sanitation, the control of communicable diseases, the protection of food, milk, and water supply, hospitals, drugs, the practice of medicine, dentistry, midwifery, pharmacy, and veterinary medicine, diseases of animals, and similar matters. The officer in charge of this department will be a doctor of medicine, preferably with training and experience as a health officer.

d. Education.—This department will supervise universities, colleges, and schools of all sorts, public and private. The officer in charge of it should have had experience in the teaching profession, preferably as an executive.

e. Public safety.—(1) This department will have as its most important responsibility the maintenance of order and prevention of crime among the civilian population. It will supervise civilian police, prisons, and fire departments, the traffic in liquor and narcotics, the circulation of civilians, identification cards, and similar matters.

(2) The military intelligence section of the general staff will establish policies, and, in consultation with this department, will prepare orders and regulations with respect to censorship. The duty of supervising their enforcement upon the inhabitants is a responsibility of this department. Though this department will not have charge of the military police, it will consult with the Provost Marshal General as to their activities insofar as they are used as an agency for the purposes above mentioned.

(3) The officer in charge of this department should have had experience in military or civilian police duties.

f. Legal.—The duties of this department will include the following and similar matters:

8

(1) Supervision of military commissions and provost courts, examination of their records and advice with respect to action thereon, and filing of such records.

(2) Supervision of the civil courts, of public prosecutors, and of the practice of law.

(3) Legal advice to the commanding general, the officer in charge of civil affairs, and other personnel engaged in military government in respect to the operation of such government and matters concerning it.

(4) The consideration of claims of inhabitants of the occupied territory against the United States or its officers, enlisted men, or employees; and of claims by the United States, its officers, enlisted men, or employees against the enemy country, its states, provinces, counties, cities, communes, or inhabitants. The officer in charge of this department should be a lawyer by profession, preferably one experienced in governmental or municipal matters.

g. Communication.—This department will supervise the postal service, telegraphs, telephones, radio, and other means of communication among the inhabitants of the occupied territory and between such inhabitants and other countries and territories. Though this department will not supervise censorship, its personnel will cooperate in its enforcement. The officer in charge of this department should have had experience in the postal service or in connection with telegraphs, telephones, or the radio.

h. Public welfare.—This department will supervise the care of the poor, infants, children, and the aged, and all charitable institutions and organizations. The officer in charge of it should have had experience as a welfare worker or in charitable institutions.

i. Economics.—This department will supervise the agriculture, manufactures, and trade of the occupied territory, its mines and oil wells, exports and imports, the supply of the inhabitants with food, fuel, and other necessaries, the supply of labor, strikes, lock-outs, and disputes, and like matters. The officer in charge of this department should have had experience in the matters with which it deals.

■ **14. Uniform Worn by Personnel on Civil Affairs Duty.**— All persons on civil affairs duty will wear the uniform of their

9

arm or service and insignia of grade and in addition a purple arm band bearing the letters "CA" in white, or such other distinctive device as the commanding general of the theater of operations may direct.

SECTION IV

ORGANIZATION

■ 15. GENERAL.—The officer in charge of civil affairs on the staff of the commanding general, theater of operations, will submit to the commanding general timely recommendations as to the organization of military government, which will be such as the commanding general may direct. The organization will depend upon the tactical situation, the geography and civil government of the occupied territory, the extent to which reliance may be placed upon its officers to remain at their posts and perform their duties satisfactorily, the characteristics and disposition of the people, and other attendant circumstances. If hostilities are still in progress, or, even if they are suspended, if there is any likelihood of their resumption, pursuant to the policy of separate personnel for military government (par. 10b), tactical units should not be made organs of military government, nor should any duties with respect to military government be imposed upon their commanding generals or officers, staffs, or personnel.

■ 16. STATES OR PROVINCES.—Pursuant to the policy of retention of existing political divisions (par. 10e), if the occupied territory is divided into states or provinces, it will generally be advisable to detail an officer in charge of civil affairs for each state or province occupied, with station at its capital, and to furnish him with appropriate commissioned, warrant, and enlisted assistants. His office should be organized similarly to that of the civil affairs section at headquarters of the theater of operations, though so many departments may not be needed.

■ 17. MILITARY DISTRICTS.—If the occupied territory is not divided into states or provinces, but has been or shall be divided by the commanding general, theater of operations, for administrative purposes into districts or sections, as was done

10

by the commander in chief, American Expeditionary Forces in France, the commanding general of each such district or section may be directed to assume responsibility for military government therein, to detail an officer in charge of civil affairs, and to set up a civil affairs section of his staff, which should be organized as above stated (par. 13).

■ 18. TACTICAL UNITS AS ORGANS OF MILITARY GOVERNMENT.— *a. Armies, corps, and divisions.*—If hostilities have ceased or if they have been suspended and the probability of their resumption is remote, armies, corps, and divisions may be used as organs of military government, and the commanding general of each army, corps, or division may be directed to assume responsibility for military government in the area occupied by his unit, to detail an officer in charge of civil affairs, and to set up a civil affairs section of his staff.

b. Smaller tactical units.—Upon the same conditions and in a similar manner, smaller tactical units may be used as organs of military government, but in general the smaller the unit the smaller is its staff, the more desirable it is that the commanding officers and staff devote their time and energy solely to the training of their men, and the greater is the risk that it will not function satisfactorily as an organ of military government.

c. Staffs.—Whenever any tactical unit is used as an organ of military government, the staff of the commanding general or officer should be organized, insofar as concerns military government, like that of the commanding general of the theater of operations (par. 13), except that in general the smaller the tactical unit the smaller the civil affairs section will be and the fewer the number of its departments. Pursuant to the basic policy of permanence (par. 9e), whenever a tactical unit used as an organ of military government is moved, its civil affairs section should be transferred to the staff of the new unit occupying the same area, or otherwise retained on the same duties.

d. Boundaries of areas.—Pursuant to the policy of retention of existing political divisions (par. 10e), the boundaries of the area occupied by a unit used as an organ of military government should coincide with existing political boundaries.

■ 19. COUNTIES AND CITIES.—In most countries there will be found a political division of approximately the size of an American county, called county, department, *Kreis,* or by some other name. It will generally be found advisable to detail an officer to supervise the government of each such division. It will also generally be found advisable to detail an officer to supervise the government of each incorporated city, except very small ones. Such an officer may be called the officer in charge of civil affairs for the county or the city, or by a similar title. Such assistants, if any, as the nature and volume of the work require will be furnished him.

SECTION V

MILITARY TRIBUNALS

■ 20. GENERAL.—Military tribunals will be of the kind, number, and composition, shall have such jurisdiction and powers, shall follow such procedure, and keep such records, as the commanding general, theater of operations may direct. Experience in past military occupations by the Army of the United States has shown such an organization as the following to be advisable, but it may be modified as the circumstances of the particular occupation may require.

■ 21. ESTABLISHMENT.—As soon as practicable after entry into the enemy's territory, the commanding general of the theater of operations will establish necessary military tribunals, and by publication of ordinances (sec. VIII) or other appropriate method will notify the inhabitants thereof of the offenses for which they may be tried by such tribunals, and of the punishments which such tribunals may impose.

■ 22. KINDS.—In general, there will be three kinds of military tribunals—
> Military commissions.
> Superior provost courts.
> Inferior provost courts.

Special tribunals may be established for the trial of vagrants, prostitutes, juveniles, or other classes of offenders, or for civil cases (par. 32).

12

■ 23. COMPOSITION.—*a.* A *military commission* shall consist of any number of officers not less than five, one of whom shall be the law member, a trial judge advocate, and a defense counsel. One or more assistant trial judge advocates and assistant defense counsel may be appointed.

b. A superior provost court shall consist of one officer, who shall be an officer of field grade unless none such is available.

c. An inferior provost court shall consist of one officer.

■ 24. BY WHOM APPOINTED.—*a. Military commissions* may be appointed by the commanding general of the theater of operations only, and, if that power shall be delegated by him to them, by the commanding generals of armies, corps, divisions, or military districts, or by the officers in charge of civil affairs for states, provinces, or military districts.

b. Superior provost courts may be appointed by the commanding general of the theater of operations, by the commanding generals of armies, corps, or divisions if those units are used as organs of military government, and by the officers in charge of civil affairs for states, provinces, or military districts.

c. Inferior provost courts may be appointed as are superior provost courts and also by the commanding officers of smaller tactical units used as organs of military government and by officers in charge of civil affairs for counties or cities. If but a single officer is on duty in a tactical unit used as an organ of military government or if but a single officer is on duty in connection with civil affairs in a county or city, such officer shall be an inferior provost court.

■ 25. JURISDICTION.—*a. Over persons.*—The military tribunals herein enumerated shall have jurisdiction over all persons within the occupied territories except those having diplomatic immunity and those subject to the military or naval law of the United States or of countries allied or associated with the United States. Persons subject to the military law of the United States charged with offenses will be tried by court martial.

b. Over offenses.—The military tribunals herein enumerated have jurisdiction over all acts or omissions made crimes or offenses by the laws of the country occupied, over offenses against the laws of war, and over violations of the proclamations, ordinances, regulations, or orders promulgated by the commanding general, theater of operations, or by any of his subordinates within the scope of his authority. However, if the courts of the occupied country are open and functioning satisfactorily, they should be permitted to try persons charged with offenses against the laws of that country not involving the United States, its property, rights, or interests, or the person, property, or rights of a member of the occupying forces. The commanding general, theater of operations, or, if that power is delegated to him, a subordinate officer in charge of civil affairs, may withdraw any case or class of cases from a court of the occupied country and direct that it or they be dropped or tried by a military tribunal. Such power should be exercised with respect to any prosecution inimical to the interests of the United States. Each military tribunal shall have jurisdiction over such charges only as may be referred to it for trial by the officer appointing it or his successor. If that officer shall consider charges deserving of severer punishment than can be imposed by the court to which he may refer them, he will forward them to higher authority, recommending reference to a higher tribunal. An officer who is an inferior provost court because he is the only officer present may try such charges as he or others may prefer, or may forward them to higher authority.

■ 26. BAIL.—As soon as conditions render it practicable to do so, the commanding general of the theater of operations, or, if that power is delegated by him to them, subordinate commanders or officers in charge of civil affairs will issue orders announcing in what cases, under what conditions, and by whom persons awaiting trial by military tribunals may be admitted to bail or released without bail but with a summons to appear for trial. Admission to bail and release without bail but with summons are matters of discretion and not of right.

■ 27. PROCEDURE.—*a. Military commissions.*—The procedure of military commissions shall be the same as that of general courts martial, except insofar as obviously inapplicable.

b. Provost courts.—The procedure of provost courts shall be the same as that of summary courts martial, except insofar as obviously inapplicable.

14

c. *Counsel.*—Every defendant before a military tribunal is entitled as a matter of right to counsel of his own selection and at his own expense, but military counsel will be provided only before military commissions.

d. *Attendance of witnesses.*—(1) *Military witnesses.*—The attendance of military witnesses will be obtained as in the case of military witnesses before courts martial.

(2) *Civilian witnesses.*—Military tribunals are authorized to compel the attendance of civilian witnesses whose testimony they may think needed or desirable. If necessary, they may require the assistance of the local civilian authorities or request that of the military police or of any appropriate commanding officer.

e. *Translation.*—If the defendant and his counsel, if any, understand and speak English, the proceedings will be conducted in that language and no interpreter will be necessary. If the personnel of the military tribunal have sufficient knowledge of that language, the proceedings may be conducted in the language of the occupied territory. If neither of these conditions exists, an interpreter will be employed who will take care that defendant, his counsel, and the personnel of the tribunal are fully informed as to the entire proceeding.

f. *Previous convictions.*—A military tribunal may consider, after a finding of guilty and before imposition of sentence, evidence of previous convictions and sentences by either military tribunals or civil courts; provided, that evidence of conviction of an offense legally punishable by imprisonment for more than one year (whether in fact so punished or not) shall be admissible without regard to the date of commission of such offense, but evidence of conviction of an offense not so punishable shall be admitted only if the offense was committed within one year next preceding the commission of any offense of which the defendant shall have been convicted at the trial then in progress.

■ 28. SENTENCES.—a. (1) *Military commissions.*—A military commission may impose any lawful and appropriate sentence, including death or life imprisonment.

(2) *Superior provost courts.*—The maximum sentence which a superior provost court may impose is confinement at hard labor for 6 months and a fine of $1,000, or both.

15

(3) *Inferior provost courts.*—The maximum sentence which an inferior provost court may impose is confinement at hard labor for one month and a fine of $100, or both.

NOTE.—The maximum fines mentioned above should be stated in round numbers in the money current in the occupied territory.

b. (1) *Expulsion.*—When such a punishment is appropriate and its execution practicable, a military commission or superior provost court, in lieu of or in addition to any other lawful punishment, may sentence a defendant to expulsion from occupied territory.

(2) *Confiscation.*—If a defendant shall be convicted of wrongful sale, purchase, use, or possession of any article or articles or of the wrongful operation of a place of business for the sale of such articles, a military tribunal, in lieu of or in addition to any other lawful punishment, may decree the forfeiture to the United States of such article or articles or the stock thereof in his possession or place of business.

(3) *Padlocking.*—If a defendant shall be twice convicted of the wrongful sale or gift of intoxicating liquor or a habit-forming drug (including marijuana) at a certain place, or twice convicted of other violations of regulations or orders with respect to the sale of intoxicating liquors at one place, or once convicted of the wrongful operation of a house of prostitution, a military tribunal, in addition to any other lawful punishment, may order that such place be vacated and closed for a fixed time.

c. Table of maximum punishments.—In order to prevent injustice by too great diversity in the sentences imposed by different tribunals for the same offense, the commanding general of the theater of operations should publish in the ordinances (sec. VIII), or otherwise, a table fixing maximum limits of punishment to be imposed by military tribunals for the more usual offenses. The table should include a statement of the equivalent of a fine in days of confinement for the use of military tribunals in imposing alternative sentences or in case the defendant cannot pay the fine.

d. Place and manner of confinement.—The commanding general, theater of operations, will issue orders as to the place of confinement of persons sentenced by military tribunals to imprisonment, as to the labor to be required of them, and as

16

to other details of their confinement. Unless military necessity (par. 9a), the termination of the occupation, or other cogent reasons require their removal, persons serving sentences imposed by military tribunals should be confined within the occupied territory.

■ 29. RECORDS.—a. Charges will be preferred by a person subject to military law on a printed "Charge Sheet" (App. III). If that form is not available in print, it may be typed or written in longhand, or the charge sheet in use in the Army (W. D., A. G. O. Form No. 115) may be used, disregarding such parts as are obviously inapplicable and making such additions as are necessary. No oath to the charges is necessary.

b. The records of trials by provost courts will be kept on the back of the charge sheet.

c. Military commissions will keep a separate record of the trial of each case in form as nearly as practicable like that of a general court martial.

■ 30. DISPOSITION OF FINES.—Fines, forfeited bail, proceeds of sales of confiscated property, and other receipts of military tribunals will be paid without any deduction to the nearest disbursing officer and deposited in the Treasury as miscellaneous receipts (31 U. S. C. 484; J. A. G., A. E. F., February 6, 1919; 3 Asst. Comp. Dec., France, 138.)

■ 31. APPROVAL, CONFIRMATION, AND REVIEW.—a. *Action by the appointing authority.*—(1) *Military commissions.*—No sentence of a military commission shall be carried into effect until it shall have been approved by the officer appointing the commission or his successor.

(2) *Provost courts.*—The sentence of a provost court shall be executed forthwith without awaiting action by higher authority. Nevertheless, every record of trial by a provost court shall be examined by the officer who appointed the court, or at his headquarters, or by his successor, and such officer or his successor shall have power to disapprove or vacate, in whole or in part, any findings of guilty, to mitigate, commute, remit, or vacate any sentence, in whole or in part, and he may restore the accused to all rights affected by the findings and sentence.

b. Final review.—Every record of trial by military commission and one copy of every record of trial by provost court shall be forwarded by the officer who appointed the tribunal or his successor, after he shall have examined it, to the officer in charge of civil affairs at headquarters of the theater of operations, where it shall be examined in the legal department of that office. The chief of that department shall submit through the officer in charge of civil affairs to the commanding general such recommendations, if any, as may be appropriate with respect thereto. No sentence of death shall be carried into execution until it shall have been confirmed by the commanding general of the theater of operations. When the authority competent to confirm the sentence has already acted as approving authority no additional confirmation by him is necessary. The commanding general of the theater of operations shall have power to disapprove or vacate, in whole or in part, any findings of guilty made by any military tribunal, to mitigate, commute, remit, or vacate any sentence, in whole or in part, imposed by such tribunal, and he may restore the accused to all rights affected by the findings and sentence. He may direct that the records of trial of inferior provost courts be forwarded to some other officer or officers, in whose headquarters or offices they shall receive such review as has been above directed, and he may delegate his powers with respect to the findings and sentences of such courts to such officer or officers.

■ 32. CIVIL CASES.—If the courts of the occupied country are open and functioning satisfactorily, they should be permitted to hear and determine civil suits other than those brought against members of the occupying forces, of which they have no jurisdiction. If the occupation is likely to be brief, no provision need be made for trial of civil cases even if the courts of the occupied country are not functioning. If, however, the courts of the occupied country are not functioning satisfactorily, and the welfare of the people so requires (par. 9b), the commanding general, theater of operations, may confer jurisdiction in civil cases upon military commissions and provost courts or may establish separate military tribunals for such cases, and may issue such regulations as to them and as to the execution of their judgments and decrees as he may think proper. The law to be followed in civil cases is that of the occupied country.

PHASES

■ 33. PHASES.—Military government will usually pass through successive phases. In many cases the occupation itself may be successive, and military government will be established in rear areas while active operations are in progress in forward areas. Though all three phases may not exist in every case, generally in a successful campaign in an enemy country each district thereof will pass through the following phases:

a. First phase, while fighting is going on in the district. During this phase little can be done to set up military government, but usually a proclamation will be issued to the people of the occupied territory (see sec. VII and app. IV). Nevertheless, unless all the inhabitants have been evacuated, combat units will of necessity come into contact with them and must deal with them, but should do so as little as possible and devote their attention primarily to their tactical duties. The civil affairs section of the staff of the commanding general, theater of operations, will prepare instructions to be issued to the commanding officers of tactical units to govern their actions in such dealings. In general, relations with the inhabitants during this period will be conducted through the provost marshal and military police. During this phase some members of the civil affairs section of the staff of the commanding general of the theater of operations should accompany tactical units in order to secure advance information of areas being occupied while other members are preparing plans and orders for the establishment of military government.

b. Second phase.—During this phase organized resistance has ceased in the district in question and military government is organized and operates, though peace has not been definitely and finally established between the United States and the occupied country. Not later than at the beginning of this phase, the commanding general should publish ordinances (sec. VIII and app. V) setting out what is required of the people of the occupied territory, and what it is forbidden that they should do. Also during this phase the civil affairs

19

section of the staff of the commanding general, theater of operations, should be planning for military government of any other districts that may be subsequently occupied.

c. *Third phase,* after fighting between the United States and the enemy nation has ceased, in consequence of an armistice or protocol which renders the resumption of hostilities highly improbable, of the surrender or destruction of the enemy's armed forces, or of a treaty of peace. During this period the basic policy of military necessity (par. 9a) operates with greatly diminished force, if at all. The policies adopted for the operation of military government will be affected to some extent during all phases, but especially during the third phase, by the anticipated future status of the occupied territory, i. e.:

(1) Permanent retention by the United States, as in Puerto Rico;

(2) Its erection into a new state, as in Cuba; or

(3) Its return to its former sovereign or to its own people, as in Vera Cruz in 1914 and in Germany in 1918–23.

In advance of the time when they will be needed, the Personnel Division (G–1) of the War Department General Staff will make plans for the transition from military to civil government. The civil affairs section of the staff of the commanding general, theater of operations, will make such further and more detailed plans for such transition as may be necessary and at the proper time will supervise such transition. Generally activities of the military government will be gradually curtailed during this period, and more and more of its operations taken over by the civil government until the latter assumes full control and the Army becomes merely a garrison or is withdrawn.

SECTION VII

PROCLAMATION

■ 34. TIME OF ISSUANCE.—Though such action is not legally prerequisite to the establishment of military government, the commanding general, theater of operations, at as early a date as is practicable during the first phase or at the latest at the beginning of the second phase, should issue a proclamation to the people of the occupied territory.

■ 35. FORM AND CHARACTER.—In order that the entire population may read it in full, the proclamation should be brief, should be clearly and idiomatically translated into the language of the occupied country, and published in English and in that language as promptly and as widely as practicable. Its tone should be dignified and firm, but not harsh or needlessly offensive. It should be signed by the commanding general of the theater of operations.

■ 36. CONTENTS.—The contents of the proclamation will vary according to the circumstances of the occupation. For a type proclamation, see appendix IV. In general, the proclamation will cover the following points:

a. Declaration of the occupation.

b. Purpose and policy of the occupation.

c. Supremacy of the military authority of the United States, involving—

(1) Suspension of political ties with and obligations to the enemy government.

(2) Obedience to the commanding general and other military authorities, particularly military police and personnel on civil affairs duty.

(3) Abstinence from acts or words of hostility or disrespect to the occupying forces.

d. Except insofar as they may be changed by the military authority, continuance in operation or on duty of—

(1) Local laws and regulations.

(2) Executive and judicial officers.

(3) Railroads and public utilities.

e. Assurance that the people will be protected by the occupying army in their persons, property, family rights, religion, and in the exercise of their occupations.

f. Duty of the people to continue or resume their usual occupations.

g. Statement that this proclamation will be accompanied or followed by more detailed ordinances (sec. VIII).

SECTION VIII

ORDINANCES

■ 37. TIME OF ISSUANCE.—At the same time as the publication of his proclamation to the people of the occupied territory or as soon thereafter as practicable, the commanding general, theater of operations, should issue ordinances regulating their conduct.

■ 38. FORM AND CHARACTER.—The ordinances should inform the people of the occupied territory what is required of them, what it is forbidden that they should do, of the tribunals before which they may be tried, and of the punishments which such tribunals may impose. In justice to the people of the occupied territory (par. 9b), offenses should be clearly defined. Unfamiliar technical terms should be avoided so far as possible, as well as vague and all-inclusive language, such as, "Whosoever commits any act whatsoever injurious to the American Army * * * will be punished as a military court may direct." The ordinances should be clearly and idiomatically translated into the language of the occupied country and published in English and in that language as promptly and as widely as is practicable. Ordinances may be amended or a new edition of them published as experience may show to be necessary; but frequent changes may be taken by the people of the occupied territory as indications of vacillation and weakness, and are to be avoided. In general, it is better policy to be strict at the beginning of an occupation and gradually to relax requirements, than to follow the opposite course.

■ 39. CONTENTS.—The contents of the ordinances will vary according to the mentality, laws, and customs of the people of the occupied territory, its geography and history, the strategical and tactical situation, and other attendant circumstances. For type ordinances, see appendix V. In general, the ordinances will cover the following points, or such of them as may be necessary:

a. Identity cards.

b. Circulation of personnel and vehicles on the highways, by rail, water, and otherwise.

c. Meetings, parades, speeches, songs, and music.

d. Enemy flag, national anthem, and uniform.

e. Communication by mail, telegraph, cable, telephone, radio, pigeon, and otherwise.

f. Newspapers, magazines, and books.

22

g. Photographs.

h. Manufactures and commerce.

i. Prices.

j. Arms, ammunition, and explosives.

k. Intoxicating liquors and narcotics.

l. Sanitation and public health.

m. Prostitution.

n. Taxes, contributions, supplies, and requisitions.

o. Billeting.

p. Various offenses:

(1) Injury or violence to the person of any member of the Army of the United States.

(2) Larceny, embezzlement, sale, purchase, receipt in pawn, or wrongful possession of or damage to any property of the United States or of any person belonging to the Army of the United States.

(3) Interference with troops or with any activity of the Army of the United States.

(4) Disobedience of or failure to obey a lawful order of the commanding general of the theater of operations or of any subordinate.

(5) False statement to any military personnel on a matter of official business or concern.

(6) Damage to, obstruction of, or interference with, roads, railroads, canals, wharves, waterworks, electric light and power plants or transmission lines, gas works, or the like.

(7) Spying, communication with the enemy, or aid to him.

(8) Spreading hostile propaganda.

(9) Escape from confinement.

(10) Disrespect to the United States, its government, flag, Army, or personnel.

(11) Aiding or advising anyone to do any of the things enumerated above.

q. Military tribunals, their kinds, jurisdiction, procedure, the maximum sentences which each kind of tribunal may impose, and a table of maximum punishments for the more usual offenses (par. 28c).

r. Claims, petitions, and complaints.

Notes

Introduction

1. Based on Myers-Briggs Type Inventory results; the dominant personality types in the army officer population are introverted sensors.
2. Field Manual 3-0, Operations, Headquarters, Department of the Army (February 27, 2008), p 6-1.
3. Cited in The U.S. Army and the Interagency Process, Historical Perspectives, The Proceedings of the Combat Studies Institute 2008 Military History Symposium, Kendall D. Gott, ed. (Fort Leavenworth, KS: Combat Studies Institute Press, 2008), pp 183, 184.
4. Thomas P. M. Barnett, *The Pentagon's New Map* (Berkley Books, 2004).

Chapter 1

1. Carl von Clausewitz, *On War*, edited and translated by Michael Howard and Peter Paret (Princeton, NJ: Princeton University Press, 1976).
2. Ibid., pp 29–41. Howard's discussion on the influence of Clausewitz.
3. Ibid., p 81.
4. Ibid., p 80.
5. Ibid., p 605.
6. Ibid., pp 591, 592.
7. Ibid., p 483.
8. Ibid., p 480.
9. Ibid., p 635.
10. Ibid., p 637.
11. Ibid., p 366.
12. Ibid., pp 258, 259.
13. Stanley Sandler, *Glad to See Them Come and Sorry to See Them Go: A History of U.S. Army Tactical Civil Affairs/Military Government, 1775–1991* (U.S. Army Special Operations Command History and Archives Division, undated), p 230.
14. Stuart Kinross, *Clausewitz and America: Strategic Thought and Practice from Vietnam to Iraq* (Routledge, 2008), pp 75, 76.
15. Harry Summers, *On Strategy: A Critical Analysis of the Vietnam War* (Novato, CA: Presidio Press, 1982). In the bibliography, Summers calls the Howard/

Paret translation "masterful . . . the language more readable . . . an under-standable and usable guide to modern strategy" (p 216).

16. Ibid., pp 122–24.
17. Ibid., p 21.
18. FM 100-5 (1986), p 10.
19. FM 3-0 (February 27, 2008), pp 6–8.
20. Joint Publication 5-0 (December 26, 2006), p C-3.
21. William G. Pierce and Robert C. Coon, "Understanding the Link between Center of Gravity and Mission Accomplishment," *Military Review* (May-June 2007), p 83.

Chapter 2

1. Casualty data from Micheal Clodfelter, et al., *Warfare and Armed Conflicts: Statistical Reference to Casualty and Other Figures, 1500–2000*, 2d ed. (Jefferson, NC: McFarland and Company, 2002). More detailed discussion and referencing is presented in chapters 3 and 4.
2. FM 100-5 (1973), p 3-4, states, "Massive and violent firepower is the chief ingredient of combat power. . . . The skillful commander substitutes firepower for maneuver whenever he can do so."
3. See FM 27-5, "Military Government" (1940), which is reprinted in the appendix. The manual provided the doctrine for the successful post–World War II occupations but is no longer part of American military doctrine.
4. John J. McGrath, *Boots on the Ground: Troop Density in Contingency Operations* (Fort Leavenworth, KS: Combat Studies Institute Press, 2006), pp 91–95.
5. Whites in the former Confederate states had practiced representative government from colonial times. Germany had representative bodies—for example, the Reichstag—even under the emperor and was a republic from the end of World War I until 1933. The Japanese Diet was established in 1890, and universal male suffrage was instituted in 1925. General MacArthur retained and strengthened the Diet as part of the post–World War II occupation.
6. Discussion of the widespread chaos in the Far East in the wake of the collapse of the Japanese Empire can be found in Ronald H. Spector, *In the Ruins of Empire: The Japanese Surrender and the Battle for Postwar Asia* (New York: Random House, 2007).

Chapter 3

1. Clodfelter, et al., *Warfare and Armed Conflicts*, p 66.
2. Robert A. Doughty, et al., *American Military History and the Evolution of Western Warfare* (Lexington, MA: D. C. Heath and Co., 1996), pp 2–5.
3. William L. Shea, "Virginia at War, 1644–1646," *Military Affairs*, Vol. 41, No. 3 (October 1977), pp 142–47.

4. "Jamestown: The Real Story," *National Geographic* (May 2007), pp 47–49.

5. Doughty, *American Military History and the Evolution of Western Warfare*, pp 5–8.

6. *American Military History*, Maurice Matloff, general ed. (Office of the Chief of Military History, U.S. Army, 1973 revised edition), pp 134–37.

7. Joseph A. Stout, Jr., "The United States and the Native Americans," cited in *The American Military Tradition from Colonial Times to the Present*, John M. Carroll and Colin F. Baxter, eds. (Lanham, MD: SR Books, 1993), p 98.

8. Andrew J. Birtle, *U.S. Army Counterinsurgency and Contingency Operations Doctrine, 1860–1941* (Washington, DC: U.S. Army Center of Military History, 2004), pp 67–69.

9. Ibid., pp 76–85.

10. Demographic data from Clodfelter, et al., *Warfare and Armed Conflicts*, pp 272–74.

11. Charles Erskine Scott Wood, "The Pursuit and Capture of Chief Joseph," www.pbs.org/weta/thewest/resources/archives/six/joseph.htm, retrieved September 4, 2007.

12. Barbara W. Tuchman, *The March of Folly* (New York: Alfred A. Knopf, 1984), pp 21–24.

13. Clodfelter, et al., *Warfare and Armed Conflicts*, pp 146, 147.

14. Wallace Brown, *The Good Americans: The Loyalists in the American Revolution* (New York: William Morrow and Company, 1969), pp 191, 192.

15. William E. Daugherty and Marshall Andrews, *A Review of U.S. Historical Experience with Civil Affairs, 1776–1954* (Bethesda, MD: Johns Hopkins University Operations Research Office, 1961), pp 15–19.

16. Ibid., p 22.

17. Sandler, *Glad to See Them Come*, pp 1, 2.

18. Doughty, *American Military History and the Evolution of Western Warfare*, p 77.

19. Retrieved at http://en.citizendium.org/wiki/California,_history_to_1845, May 14, 2008.

20. Daugherty and Andrews, *U.S. Historical Experience with Civil Affairs*, p 47.

21. Doughty, *American Military History and the Evolution of Western Warfare*, pp 82–84.

22. K. Jack Bauer, *Mexican War: 1846–1848* (Lincoln, NE: University of Nebraska Press, 1974), pp 12, 13, 128, 171, 172, 193.

23. The Battle of Los Angeles in January 1847 is illustrative of the halfhearted military resistance. An army of some four hundred Californios, perhaps half the military-age Hispanic males in the area, quit the field and dispersed to their homes after conducting two demonstration attacks against a six-hundred-man American invasion force. Total Californio losses in the two

engagements amounted to only two killed and a handful of wounded and injured. Ibid., pp 189–92.

24. Daugherty and Andrews, *U.S. Historical Experience with Civil Affairs*, pp 46–50.
25. Clodfelter, et al., *Warfare and Armed Conflicts*, p 280.
26. Ibid.
27. Birtle, *U.S. Army Counterinsurgency and Contingency Operations Doctrine, 1860–1941*, p 17.
28. Daugherty and Andrews, *U.S. Historical Experience with Civil Affairs*, p 83.
29. Ibid., p 16.
30. Bauer, *Mexican War*, pp 381, 382.
31. Birtle, *U.S. Army Counterinsurgency and Contingency Operations Doctrine, 1860–1941*, p 16.
32. Daugherty and Andrews, *U.S. Historical Experience with Civil Affairs*, p 78.
33. Ibid., p 80.
34. Ibid., p 87.
35. Bauer, *Mexican War*, pp 378–87.
36. Sherman never said all these lines in this sequence in a single composition. The author assembled these lines into logical sequence. Quotations from various sources, mainly Birtle, *U.S. Army Counterinsurgency and Contingency Operations Doctrine, 1860–1941*, pp 36, 37.
37. Birtle, *U.S. Army Counterinsurgency and Contingency Operations Doctrine, 1860–1941*, p 37.
38. Lisa M. Brady, "The Wilderness of War: Nature and Strategy in the American Civil War," www.historycooperative.org/journals/eh/10.3/brady.html, paragraph 35, retrieved September 4, 2007.
39. Clodfelter, et al., *Warfare and Armed Conflicts*, pp 331–33.
40. americancivilwar.com/south/lee.html, retrieved September 7, 2007.
41. Daugherty and Andrews, *U.S. Historical Experience with Civil Affairs*, p 97.
42. Sandler, *Glad to See Them Come*, p 55.
43. Birtle, *U.S. Army Counterinsurgency and Contingency Operations Doctrine, 1860–1941*, pp 32–36.
44. General Orders 100 (Section I, 4), cited in Sandler, *Glad to See Them Come*, p 54.
45. Daugherty and Andrews, *U.S. Historical Experience with Civil Affairs*, pp 104–9.
46. James E. Sefton, *The United States Army and Reconstruction, 1865–1877* (Baton Rouge, LA: Louisiana State University Press, 1967), pp 5–7.
47. Ibid., pp 109–13.
48. Ibid., pp 261, 262.

49. Birtle, *U.S. Army Counterinsurgency and Contingency Operations Doctrine, 1860–1941*, p 57.

50. H. W. Wilson, *Downfall of Spain: Naval History of the Spanish American War* (Boston, MA: Little, Brown and Company, 1900), p 65.

51. Stephen Kinzer, *Overthrow: America's Century of Regime Change from Hawaii to Iraq* (New York: Times Books, 2006), p 41.

52. David M. Edelstein, *Occupational Hazards: Success and Failure in Military Occupation* (Ithaca, NY: Cornell University Press, 2008).

53. Birtle, *U.S. Army Counterinsurgency and Contingency Operations Doctrine, 1860–1941*, pp 104–6.

54. Ibid., pp 168–74.

55. Stuart Creighton Miller, *Benevolent Assimilation: The American Conquest of the Philippines, 1899–1903* (New Haven, CT: Yale University Press, 1982), p 94.

56. Birtle, *U.S. Army Counterinsurgency and Contingency Operations Doctrine, 1860–1941*, p 130.

57. All data in this paragraph from Robert D. Ramsey III, *A Masterpiece of Counter Guerrilla Warfare: BG J. Franklin Bell in the Philippines, 1901–1902* (Fort Leavenworth, KS: Combat Studies Institute Press, 2007).

58. Robert D. Ramsey III, *Savage Wars of Peace: Case Studies of Pacification in the Philippines, 1900–1902* (Fort Leavenworth, KS: Combat Studies Institute Press, 2007), p 102.

59. Ibid., pp 116, 117.

60. Clodfelter, *Warfare and Armed Conflicts*, p 272.

61. Ramsey, *Savage Wars of Peace*, p 57.

62. Kinzer, *America's Century of Regime Change*, p 94.

63. All World War I mobilization and casualty data from Clodfelter, et al., *Warfare and Armed Conflicts*, pp 479–84.

64. Clodfelter, et al., *Warfare and Armed Conflicts*, pp 581–84; POW number from Lucius Clay, *Decision in Germany* (New York: Doubleday, 1950), p 15.

65. www.ibiblio.org/pha/policy/1945/450508b.html, retrieved September 4, 2007.

66. Clay, *Decision in Germany*, pp 55, 56.

67. FM 27-5, Military Government and Civil Affairs (1940), p 5.

68. Clay, *Decision in Germany*, pp 65, 66.

69. Earl F. Ziemke, *The U.S. Army in the Occupation of Germany, 1944–1946* (Washington, DC: Center of Military History, United States Army, 1990), p 7.

70. Clay, *Decision in Germany*, p 9.

71. Harry L. Coles, *Civil Affairs: Soldiers Become Governors* (Washington, DC: Office of the Chief of Military History, Department of the Army, 1964), p 675.

72. Ziemke, *U.S. Army in the Occupation of Germany*, pp 134, 135.

73. Ibid., p 84.

74. Daugherty and Andrews, *U.S. Historical Experience with Civil Affairs*, p 296.

75. McGrath, *Boots on the Ground*, p 17.

76. Clay, *Decision in Germany*, p 67.

77. Ibid., pp 85–87.

78. Ibid., pp 256, 257.

79. Daugherty and Andrews, *U.S. Historical Experience with Civil Affairs*, p 299.

80. Clay, *Decision in Germany*, pp 246–49.

81. Sandler, *Glad to See Them Come*, p 234.

82. Clay, *Decision in Germany*, pp 84–90.

83. Ibid., p 268.

84. Coles, *Civil Affairs*, pp 667–69.

85. *After-Action Report, Third U.S. Army, 1 Aug. 1944–9 May 1945*, cited in Merritt Y. Hughes, "Civil Affairs in France," in *American Experiences in Military Government in World War II*, Carl J. Friedrich and Associates (New York: Rinehart and Company, 1948), p 148.

86. Ibid.

87. Ibid., pp 148, 149.

88. Ted Rall, "Dubious Liberators: Allied Plans to Occupy France, 1942–1944," www.rall.com/longarticle_11, retrieved January 2, 2008.

89. Ibid.

90. Hughes, pp 149, 150.

91. William Manchester, *American Caesar* (Boston, MA: Little, Brown and Company, 1978), p 465.

92. Clodfelter, et al., *Warfare and Armed Conflicts*, pp 581–84.

93. John W. Dover, *Embracing Defeat: Japan in the Wake of World War II* (New York: W. W. Norton and Company/The New Press, 1999), pp 47–49.

94. Douglas MacArthur, *Reminiscences* (New York: McGraw-Hill, 1964), p 281.

95. www.ibiblio.org/hyperwar/PTO/Dip/Crane.html, retrieved September 4, 2007.

96. Manchester, *American Caesar*, p 466.

97. Daugherty and Andrews, *U.S. Historical Experience with Civil Affairs*, p 383.

98. Sandler, *Glad to See Them Come*, p 269.

99. Marlene J. Mayo, "American Wartime Planning for Occupied Japan: The Role of the Experts" in *Americans as Proconsuls: United States Military Government in Germany and Japan, 1944–1952,* Robert Wolfe, ed. (Carbondale, IL: Southern Illinois University Press, 1984), p 7.

100. Ibid., pp 28, 29.

101. Ibid., p 14.

102. Ibid., pp 30, 31.

103. Ibid., pp 45–47.

104. Ibid., p 470.

105. Arthur D. Bouterse, et al., "American Military Government Experience in Japan," in *American Experiences in Military Government in World War II*, Carl J. Friedrich (New York: Rinehart and Company, 1948), pp 320, 321.

106. Lawrence A. Yates, *The U.S. Military's Experience in Stability Operations, 1789–2005* (Fort Leavenworth, KS: Combat Studies Institute Press, 2006), p 77.

107. Manchester, *American Caesar*, pp 466, 467.

108. Robert C. Orr, *Winning the Peace: An American Strategy for Post-Conflict Reconstruction* (Washington, DC: CSIS Press, 2004), pp 174–82.

109. Manchester, *American Caesar*, p 465.

110. Ibid., p 474.

111. McGrath, *Boots on the Ground*, pp 28–31.

112. Dover, *Embracing Defeat*, pp 548, 549.

113. All raw data in this paragraph from Clodfelter, et al., *Warfare and Armed Conflicts*, p 582.

Chapter 4

1. George Kennan, "X Article" (The Sources of Soviet Conduct), *Foreign Affairs*, Vol. XXV (July 1947), pp 566–82.

2. Daugherty and Andrews, *U.S. Historical Experience with Civil Affairs*, pp 396, 397.

3. Ibid., p 398.

4. Sandler, G*lad to See Them Come*, p 317.

5. Ibid., pp 312, 316.

6. Ibid., p 319.

7. Ibid., p 324.

8. Andrew J. Birtle, *U.S. Army Counterinsurgency and Contingency Operations Doctrine, 1942–1976* (Washington, DC: U.S. Army Center for Military History, 2006), p 86.

9. Ibid., pp 90–94.

10. Sandler, *Glad to See Them Come*, p 330.

11. Office of the Provost Marshal General, Department of the Army, "Activities Relating to the Korean Conflict, 25 June 1950–8 September 1951" (undated), p 8.

12. Clodfelter, et al., *Warfare and Armed Conflicts*, p 734.

13. Birtle, *U.S. Army Counterinsurgency and Contingency Operations Doctrine, 1942–1976*, p 117.

14. Stanley Karnow, *Vietnam: A History* (London: Penguin, 1984), p 137.

15. Ibid., p 146.
16. Konrad Kellen, a RAND Corporation expert who interrogated North Vietnamese and Viet Cong prisoners, cited in Karnow, *Vietnam: A History*, p 460.
17. Karnow, *Vietnam: A History*, pp 217, 218.
18. Kinzer, *America's Century of Regime Change*, p 153.
19. Karnow, *Vietnam: A History*, p 238.
20. Mobilization and casualty data from Clodfelter, et al., *Warfare and Armed Conflicts*, pp 786–91.
21. Karnow, *Vietnam: A History*, p 642.
22. Ibid., p 169.
23. Sandler, *Glad to See Them Come*, p 356.
24. Thomas W. Scoville, *Reorganizing for Pacification Support* (Washington, DC: U.S. Army, Center of Military History, 1982), pp 3–15.
25. Ibid., p 47.
26. Ibid., p 67.
27. Robert M. Perito, "The U.S. Experience with Provincial Reconstruction Teams in Afghanistan: Lessons Identified" (Washington, DC: United States Institute of Peace, October 2005), p 14.
28. Sandler, *Glad to See Them Come*, p 364.
29. David Miller, *Cold War: A Military History* (New York: St. Martin's Press, 1998), p 359.
30. Harriet Fast Scott and William F. Scott, *Armed Forces of the USSR*, 3rd ed. (Oak Forest, IL: Westview Press, 1984), pp 152, 154, 158, 245.
31. Miller, *Cold War: A Military History*, pp 250–56.
32. Matloff, *American Military History*, pp 538–42; Doughty, *American Military History and the Evolution of Western Warfare*, pp 581, 582.
33. General William DePuy, "Generals Black and Von Mellenthin: On Tactics: Implications for Nato Military Doctrine" (BDM Corporation, December 1980).
34. GlobalSecurity.org, http://www.globalsecurity.org/military/world/russia/mo-budget.htm, retrieved August 23, 2007.
35. Scott, *Armed Forces of the USSR*, p 302.
36. Clodfelter, et al., *Warfare and Armed Conflicts*, p 713.
37. Robert A. Pastor, *Whirlpool: U.S. Foreign Policy toward Latin America and the Caribbean* (Princeton, NJ: Princeton University Press, 1992), p 90.
38. Kinzer, *America's Century of Regime Change*, pp 250–54.
39. Clodfelter, et al., *Warfare and Armed Conflicts*, p 714.
40. Pastor, *Whirlpool*, pp 92, 93.

Chapter 5

1. Clodfelter, et al., *Warfare and Armed Conflicts*, pp 236–38.
2. Ibid., pp 697–99.
3. Yan Hoa, "Tibetan Population in China: Myths and Facts Reexamined," *Asian Ethnicity* (March 2000), p 15.
4. Ibid., p 25.
5. Ibid., p 22.
6. Clodfelter, et al., *Warfare and Armed Conflicts*, pp 223, 224.
7. Ibid., pp 607–10.
8. Jane's Information Group, http://www8.janes.com, Jane's Sentinel Security Assessment—Russia and the CIS, Chechen Republic, retrieved August 28, 2007.
9. Fred Weir, "Chechnya: Russia Declares 'Mission Accomplished' in Strong-man State," *Christian Science Monitor*, April 17, 2009, www.csmonitor. com/2009/0417/p06s07-woeu.html?page=1, retrieved July 16, 2009.

Chapter 6

1. Susan L. Woodward, *Balkan Tragedy: Chaos and Dissolution after the Cold War* (Washington, DC: Brookings Institute, 1995), pp 89–91.
2. Clodfelter, et al., *Warfare and Armed Conflicts*, p 604.
3. Woodward, *Balkan Tragedy*, pp 183–89.
4. Ibid., p 45.
5. Clodfelter, et al., *Warfare and Armed Conflicts*, p 604.
6. Deborah D. Avant, *The Market for Force: The Consequences of Privatization* (Cambridge, England: Cambridge University Press, 2005), pp 101–6.
7. Woodward, *Balkan Tragedy*, p 33.
8. Ibid., pp 180, 181.
9. Noel Malcolm, *Bosnia: A Short History* (New York: New York University Press, 1994), p 227.
10. Woodward, *Balkan Tragedy*, pp 172, 194.
11. John A. Tirpak, "Deliberate Force," *Air Force Magazine* (October 1997), p 40.
12. Ibid., p 41.
13. Ibid., pp 41–43.
14. Clodfelter, et al., *Warfare and Armed Conflicts*, p 605.
15. Max Boot, *Savage War of Peace* (New York: Basic Books, 2002), p 325.
16. Woodward, *Balkan Tragedy*, p 34.
17. Tim Judah, *Kosovo: War and Revenge* (New Haven, CT: Yale University Press, 2000), p 53.
18. Malcolm, *Bosnia: A Short History*, pp 212, 223.
19. Judah, *Kosovo: War and Revenge*, pp 102, 103.
20. Ibid., pp 151, 152.

21. Ibid., pp 162, 163.
22. Ibid., p 177.
23. Ibid., p 189.
24. Benjamin S. Lambeth, "NATO's Air War for Kosovo: A Strategic and Operational Assessment" (RAND, 2001), p 8.
25. Judah, *Kosovo: War and Revenge*, p 229.
26. Ibid., p 240.
27. Lambeth, "NATO's Air War for Kosovo," p 9.
28. Judah, *Kosovo: War and Revenge*, p 241.
29. Lambeth, "NATO's Air War for Kosovo," pp 35, 50.
30. Ibid., pp 38–42.
31. Ibid., p 147.
32. Ibid., pp 150, 151.
33. Ibid., pp 49, 53–56.
34. Ibid., p iv.
35. Judah, *Kosovo: War and Revenge*, pp 263, 284.
36. Lambeth, "NATO's Air War for Kosovo," pp 73–75.
37. Ibid., p 76.
38. Clodfelter, et al., *Warfare and Armed Conflicts*, p 605.
39. Paul Richter, "Air-Only Campaign Offers a False Sense of Security, Some Say," *Los Angeles Times* (June 4, 1999).
40. Thomas E. Ricks and Anne Marie Squeo, "Kosova Campaign Showcased the Effectiveness of Air Power," *Wall Street Journal* (June 4, 1999).
41. Serge Schmemann, "Now, Onward to the Next Kosovo. If There Is One," *New York Times* (June 16, 1999).
42. "Two Cheers for Airpower," *Wall Street Journal* (June 11, 1999).
43. "Rethinking Transformation," *Armed Forces Journal International* (March 2001), p 41.
44. Makhmut Gereev and Vladimir Slipchenko, *Future War* (Foreign Military Studies Office, 2007), pp 20, 32.
45. Ibid., p 31.
46. "Inside the Army" (June 28, 1999), cited in Lambeth, "NATO's Air War for Kosovo," pp 73–75.
47. army-technology.com/projects/stryker/, retrieved May 29, 2008.
48. "Futuristic Army Vision: The Service's Future Combat System Is a True Leap-Ahead Program," *Armed Forces Journal International* (May 2001), pp 26, 28.
49. Andrew Koch, "Bush reveals vision of a mobile stealthy force," *Jane's Defense Weekly* (February 21, 2001), p 3.
50. Andrew Koch, "QDR aims to transform US forces," *Jane's Defense Weekly* (August 22, 2001), p 6.
51. Ibid.

Chapter 7

Epigraph. Thomas S. Kuhn, *The Structure of Scientific Revolutions* (Chicago, IL: University of Chicago Press, 1970), p 53.

1. Field Manual 3-0, Operations (February 27, 2008), Foreword.
2. H. Norman Schwarzkopf, *It Doesn't Take a Hero* (New York: Bantam, 1992), pp 319, 320.
3. G. Bush and B. Scowcroft, "Why We Didn't Remove Saddam Hussein," *Time* (March 2, 1998).
4. Ibid., pp 468–72.
5. *Frontline*, Oral History: Norman Schwarzkopf, http://www.pbs.org/wgbh/pages/frontline/gulf/oral/schwarzkopf/7.html, July 19, 2007.
6. Ibid.
7. Joint Publication 3-0, Joint Operations (September 17, 2006), p I-17.
8. Gregory Fontenot, *On Point: The United States Army in Operation Iraqi Freedom* (Fort Leavenworth, KS: Combat Studies Institute Press, 2004), chapter 4.
9. Tommy Franks, *American Soldier* (New York: Regan Books, 2004), p 530.
10. George Packer, *The Assassins' Gate* (New York: Farrar, Straus and Giroux, 2005), p 133.
11. Ibid., p 529.
12. Ibid., p 337.
13. Thomas E. Ricks, *Fiasco: The American Military Adventure in Iraq* (London: Penguin Press, 2006), p 79.
14. Jay Garner, cited in *Turning Victory into Success: Military Operations After the Campaign*, Brian De Toy, general ed. (Fort Leavenworth, KS: Combat Studies Institute Press, 2004), pp 253–56.
15. Packer, *Assassins' Gate*, pp 138, 139; Garner, *Turning Victory into Success*, pp 258, 259.
16. Packer, *Assassins' Gate*, p 142.
17. Ibid., p 138.
18. John Agresto, *Mugged by Reality: The Liberation of Iraq and the Failure of Good Intentions* (Lanham, MD: Encounter Books, 2007), pp 68, 21.
19. Colonel Eric Nantz, e-mails to the author (April 30, 2008).
20. Peter Bouckaert /Human Rights Watch, "Violent Response: The U.S. Army in al-Falluja" (June 2003), p 4.
21. Nantz, e-mails to the author.
22. Ricks, *Fiasco*, p 140.
23. Ibid., p 142.
24. Bouckaert, "Violent Response," p 3.
25. L. Paul Bremer III, *My Year in Iraq* (New York: Simon and Schuster, 2005), pp 5–8.

26. Packer, *Assassins' Gate*, p 190.

27. Garner, *Turning Victory into Success*, p 265.

28. Rajiv Chandrasekaran, *Imperial Life in the Emerald City: Inside Iraq's Green Zone* (New York: Alfred A. Knopf, 2007), pp 74–77.

29. Ibid., p 163.

30. Ibid., pp 246–48.

31. Task Force Modularity White Paper, *Modular Brigade Combat Teams Part III* (July 15, 2004, draft), pp 4–12; Colonel Brian G. Watson, "Reshaping the Expeditionary Army to Win Decisively: The Case for Greater Stabilization Capacity in the Modular Force" (Carlisle Barracks, PA: Strategic Studies Institute, August 2005), pp 3–6, 12.

32. Chandrasekaran, *Imperial Life in the Emerald City*, pp 260, 261.

33. I do not blame the young soldiers except in the most clear-cut cases—for instance, rape. I have two twenty-year-old boys and I know how easily they can make poor choices, especially when not provided proper leadership. The systematic failure of the army leadership to adequately prepare, supervise, and discipline the young soldiers serving in Iraq is the issue here. Except in the most extreme cases, we should absolve our young soldiers of any guilt.

34. Brigadier Nigel Aylwin-Foster, "Changing the Army for Counterinsurgency Operations," *Military Review* (November–December 2005), p 5.

35. Agresto, *Mugged by Reality*, pp 158–65.

36. John M. Broder, "State Dept. Plans Tighter Control of Security Firm," *New York Times* http://www.nytimes.com/2007/10/06/washington/06blackwater.html?_r=1&hp, retrieved February 28, 2009.

37. President Bush, Osan Air Base, November 19, 2005.

38. *Report to Congress: Measuring Stability and Security in Iraq* (October 13, 2005), pp 31, 37.

39. Ahmed Hashim, of the Naval War College, cited in "Why Iraq Has No Army," James Fallows, *Atlantic Monthly* (December 2005), p 70.

40. Field Manual 3-24, Counterinsurgency (December 15, 2006), p vii.

41. For instance, Ralph Peters, "Getting Counterinsurgency Right," *New York Post* (December 20, 2006).

42. FM 3-24, p 1-1.

43. Ibid., p 6-1.

44. Anna Badkhen, "Civilian Killings, Unending Violence Appear Unstoppable," *San Francisco Chronicle* (July 19, 2006), p 1.

45. Ralph Peters, "Last Gasps in Iraq," *USA Today* (November 2, 2006), p 13.

46. Kimberly Kagan, "How They Did It: Executing the Winning Strategy in Iraq," *Weekly Standard* (November 19, 2007), p 20.

47. Philip Shishkin, "In Baghdad Neighborhood, A Tale Of Shifting Fortunes," *Wall Street Journal* (October 31, 2007), p 1.

48. Babak Dehghanpisheh and John Barry, "The Brains Behind the Petraeus Iraq Report," *Newsweek* (September 17, 2007).

49. Associated Press, "Leaders Clash on Policy," *Kansas City Star* (July 29, 2007), p A16.

50. Michael E. O'Hanlon and Kenneth M. Pollack, "A War We Just Might Win," *New York Times* (July 30, 2007).

51. Joseph Gregoire and Timothy O'Hagan, presentation to Command and General Staff College faculty on OIF PRTs (December 18, 2007).

52. Michael E. O'Hanlon, Brookings Institute, "Iraq Index: Tracking Variables of Reconstruction and Security in Post-Saddam Iraq" (July 26, 2007), p 13.

53. *Jim Lehrer Newshour,* PBS (July 16, 2007).

54. "The Human Cost of the War in Iraq: A Mortality Study, 2002–2006," Bloomberg School of Public Health at Johns Hopkins University and the School of Medicine of Al Mustansiriya University in Baghdad (October 2006).

55. Tina Susman, "Poll: Civilian Death Toll in Iraq May Top 1 Million," *Los Angeles Times* (September 14, 2007).

56. See, for instance, Greg Jaffe, "How Courting Sheiks Slowed Violence in Iraq," *Wall Street Journal* (August 8, 2007), p 1.

57. U.S. Census Bureau International Data Base.

58. Kristele Younes, Refugees International, and Dana Graber Ladek, International Organization for Migration, on *Jim Lehrer Newshour* (July 26, 2007).

59. James Glanz and Stephen Farrell, "More Iraqis Said to Flee Since Troop Rise," *New York Times* (August 24, 2007), p 1.

60. "Prospects for Iraq's Stability: Some Security Progress but Political Reconciliation Elusive," National Intelligence Estimate (August 2007), p 1-3.

61. Steven Lee Myers and Thom Shanker, "Petraeus Urges Halt in Weighing New Cut in Force," *New York Times* (April 9, 2008), p 1.

62. Sudarsan Raghavan, "A Quiet Filled with Wariness: Many in Baghdad's Tobji District Fear Their Troubles Aren't All Past," *Washington Post* (February 26, 2009), p 1.

63. Ernesto Londono, "U.S. Troops on Edge as Rules Shift in Iraq: Americans Must Coordinate with Sometimes Unreliable Local Counterparts," *Washington Post* (January 12, 2009), p 1.

64. Agreement between the Republic of Iraq and the United States of America Regarding the Withdrawal of the American Forces from Iraq and Regulating their Activities During their Temporary Presence in It, Article 24.

65. James Risen, "KBR losing exclusive hold on Iraq contracts," *International Herald Tribune* (May 26, 2008).

66. Kris Osborn, "Iraq Plans to Buy 2,000 Tanks: U.S. Firm Would Rebuild East European T-72s," *Defense News* (January 12, 2009), p 1.

67. Peter Baker and Thom Shanker, "Obama's Iraq Plan Has December Elections as Turning Point for Pullout," *New York Times* (February 26, 2009), p 1.
68. Richard Perle, et al., "A Clean Break: A New Strategy for Securing the Realm" (Jerusalem and Washington: Institute for Advanced Strategic and Political Studies, 2006).
69. ABC News/Washington Post Poll, April 10–13, 2008; Associated Press-Ipsos Poll, April 7–9, 2008; CBS News Poll, March 15–18, 2008; CNN/Opinion Research Corporation Poll, March 14–16, 2008.

Chapter 8

1. Barnett R. Rubin, "Saving Afghanistan," *Foreign Affairs* (January/February 2007), p 66.
2. General Dan K. McNeill, address to Land Forces Symposium in Islamabad, Pakistan, April 13, 2007, reprinted by Strategic Studies Institute, pp 3, 4.
3. Steve Coll, *Ghost Wars* (London: Penguin Books, 2004), p 281.
4. Ibid., pp 55–65.
5. Ibid., pp 67, 68.
6. Ibid., pp 86, 87, 180.
7. Clodfelter, et al., *Warfare and Armed Conflicts*, p 671.
8. Coll, *Ghost Wars*, p 89.
9. Ibid., pp 102, 151.
10. Ibid., p 152.
11. Ibid., p 159.
12. Ibid., p 133.
13. Kenneth Katzman, "Afghanistan: Post-War Governance, Security, and U.S. Policy," Congressional Research Service Report for Congress (January 11, 2007), p 39.
14. Clodfelter, et al., *Warfare and Armed Conflicts*, p 672.
15. Coll, *Ghost Wars*, p 172.
16. Ibid., pp 216, 217, 232, 233.
17. Ibid., p 239.
18. Ibid., p 263.
19. Katzman, "Afghanistan: Post-War Governance," pp 6, 7.
20. Sean Naylor, *Not a Good Day to Die* (Berkley Books, 2005), pp 17–21.
21. Katzman, "Afghanistan: Post-War Governance," pp 7–9.
22. Ibid., p 9.
23. Rubin, "Saving Afghanistan," p 66.
24. Perito, "U.S. Experience with Provincial Reconstruction Teams," pp 1–5.
25. Ibid., p 10.
26. Michael J. Dziednic and Colonel Michael K. Seidl, "Provincial Reconstruction Teams: Military Relations with International and Nongovernmental

Organizations in Afghanistan (Washington, DC: United States Institute of Peace, August 2005), p 5.

27. Perito, "U.S. Experience with Provincial Reconstruction Teams," p 5.

28. Katzman, "Afghanistan: Post-War Governance," pp 35–38.

29. Arnaud de Borchgrave, "Middle Ages Redux?" *Washington Times* (July 5, 2006), p 14.

30. Rubin, "Saving Afghanistan," p 8.

31. Katzman, "Afghanistan: Post-War Governance," p 30.

32. Rubin, "Saving Afghanistan," p 8.

33. Griff Witte and Imtiaz Ali, "47 Killed As Insurgents Take Key Fort in NW Pakistan," *Washington Post* (January 17, 2008), p 14.

34. Ahmed Rashid, "Pakistan's Deal with the Devil," *Los Angeles Times* (February 24, 2009), p 15.

35. Katzman, "Afghanistan: Post-War Governance," p 29.

36. Ibid., p 29.

37. Cited in Borchgrave, "Middle Ages Redux?"

38. Yochi J. Dreazen, "Britain Sees Role for Afghan Tribes," *Wall Street Journal* (January 10, 2008), p 4.

39. Helene Cooper, "Putting Stamp on Afghan War, Obama Will Send 17,000 Troops, *New York Times* (February 18, 2009), p 1.

40. Jon Cohen and Jennifer Agiesta, "Poll of Afghans Shows Drop in Support for U.S. Mission," *Washington Post* (February 10, 2009), p 11.

41. Ralph Peters, "The Mendacity of Hope," *USA Today* (February 24, 2009), p 11.

42. McNeill, address to Land Forces Symposium, p 6.

Chapter 9

1. Russell F. Weigley, *The American Way of War: A History of United States Military Policy and Strategy*, 2nd ed. (Bloomington, IN: Indiana University Press, 1977), p xx.

2. FM 100-5, Operations (1944), pp 32, 109.

3. George S. Patton, Jr., Letters of Instruction to the Third U.S. Army, March 6, 1944, cited in *War As I Knew It* (New York: Bantam, 1980), pp 379, 380.

4. Patton's data showed that his battle casualties were less than one-tenth the opposing German battle casualties, truly remarkable given that Patton's army was on the attack. Ibid., p 314.

5. FM 100-5 (1973), pp 3-4, 3-5.

6. Joint Publication 3-0: Joint Operations (November 17, 2006), p I-12.

7. Ibid., p I-14.

8. Ibid., pp I-15 to I-17.

9. Ibid., p V-1.

10. Ibid., p V-2.

11. Ibid., p V-23.
12. FM 1, The Army (June 2005), p 3-6.
13. FM 3-0 (February 27, 2008), p 3-1.
14. Ibid., p 3-7.
15. Ibid., p 3-12.
16. FM 1-02, Operational Terms and Graphics (September 2004).
17. Birtle, *U.S. Army Counterinsurgency and Contingency Operations Doctrine, 1942–1976*, pp 251, 454.
18. Ibid., p 482.
19. FM 3-0, p 2-5.
20. Stuart Kinross, *Clausewitz and America: Strategic Thought and Practice from Vietnam to Iraq* (Routledge, 2008), p 10.
21. For instance, Ralph Peters, "Getting Counterinsurgency Right," *New York Post* (December 20, 2006).
22. FM 3-24, Counterinsurgency (December 2006), p 1-1.
23. Ibid., p 1-2.
24. Ibid., p 1-19.
25. FM 3-0, p 1-15.

Chapter 10
1. Barnett, *The Pentagon's New Map*, p 25.
2. Ibid., pp 299–303.
3. Ibid., p 177, 186, 303.
4. Thomas P. M. Barnett, *Blueprint for Action* (New York: G. P. Putnam's Sons, 2005), p 39.
5. Ibid., p 40.
6. Ibid., pp 32, 33.
7. Ibid., p 30.
8. Ibid., p 52.
9. *Transforming for Stabilization and Reconstruction Operations*, Center for Technology and National Security Policy, National Defense University, edited by Hans Binnendijk and Stuart E. Johnson (Washington, DC: NDU Press, 2004), pp 55–61.
10. James B. Ellsworth, "SysAdmin: Toward Barnett's Stabilization and Reconstruction Force," Land Warfare Paper No. 57, Association of the United States Army, 2006.
11. Ibid., p v.
12. Ibid., pp 9, 10.
13. Sarah Sewell, quoted in Samantha Power, "Our War on Terror," *New York Times Sunday Book Review* (July 29, 2007).
14. Michael E. O'Hanlon, "Iraq Index," p 6.

15. Dane F. Smith, Jr., "Roundtable on Proposed Civil Reserve Corps," Center for Strategic and International Studies, November 2007, pp 1, 2.
16. See, for instance, FM 3-24, p 5-3, or FM 3-0 (2004 draft), p 6-14.
17. See, for instance, Coles, *Civil Affairs*, pp 3–6.
18. See, for instance, Thomas E. Ricks, "The Lessons of Counterinsurgency: U.S. Unit Praised for Tactics Against Iraqi Fighters, Treatment of Detainees," *Washington Post* (February 16, 2006), p 14.
19. The number is only an estimate. The European Civil Affairs Division, organized on January 13, 1944, comprised 8,263 soldiers. Coles, *Civil Affairs*, p 675.
20. FM 27-5, "Military Government" (1940), p 5.
21. Ibid.
22. See, for instance, David Galula, *Counterinsurgency Warfare: Theory and Practice* (St. Petersburg, FL: Hailer Publishing, 2005), pp 127–33.
23. Sandler, *Glad to See Them Come*, p 364.
24. Vichy France is the noteworthy exception.
25. *Measuring Stability and Security in Iraq*, DoD Report to Congress (June 2007), p 30.
26. John A. Nagl, "Institutionalizing Adaptation: It's Time for a Permanent Army Advising Corps" (Washington, DC: Center for a New American Security, June 2007), p 5.
27. Yaweri Museveni, former insurgent commander and current president of Uganda, speech to CGSC class, September 26, 2008.

Chapter 11

1. Daugherty and Andrews, *U.S. Historical Experience with Civil Affairs*, p 77.
2. Ibid., p 200.
3. Ibid., pp 301, 302.
4. Kinzer, *Overthrow: America's Century of Regime Change*, pp 202, 203.
5. Joseph E. Stiglitz, *Making Globalization Work* (New York: W. W. Norton & Co., 2006), pp 37–40.
6. Andrew J. Birtle, *U.S. Army Counterinsurgency and Contingency Operations Doctrine, 1860–1941*, p 226.
7. Spector, *In the Ruins of Empire*, pp 48, 242–54.
8. William R. Swarm, "Impact of the Proconsular Experience on Civil Affairs Organization and Doctrine," in *Americans as Proconsuls: United States Military Government in Germany and Japan, 1944–1952*, Robert Wolfe, ed. (Carbondale, IL: Southern Illinois University Press, 1984), p 412.
9. Birtle, *U.S. Army Counterinsurgency and Contingency Operations Doctrine, 1942–1976*, p 216.
10. Ibid., p 233.

11. Chronology from http://www.globalspecialoperations.com/usasoc.html, August 27, 2007.
12. Swarm, "Impact of the Proconsular Experience," pp 413, 414.
13. Chronology from http://www.globalspecialoperations.com/usasoc.html, August 27, 2007.

Chapter 12

1. Michael A. Ledeen, "Syria and Iran Must Get Their Turn," *National Post*, Canada (April 7, 2003).
2. Robert Baer, "Is a U.S.-Iran War Inevitable?" *Time* (March 29, 2007).
3. Birtle, *U.S. Army Counterinsurgency and Contingency Operations Doctrine, 1860–1941*, p 197. Asked by President Taft to consider an invasion of Mexico, where growing instability threatened U.S. border settlements and anti-American revolutionaries, some with their own armies, flourished, the army estimated that it would need 400,000 to 550,000 men and three to four years to pacify Mexico, a nation of only 15 million people. The army estimated that the initial invasion would take only a few months but that the counterguerilla and nation-building operations, based on the recent experience in the Philippines and South Africa, would be the truly difficult part of the effort. President Taft decided to deal with Mexico in other ways.
4. Clodfeter, et al., *Warfare and Armed Conflicts*, pp 617, 618.
5. William Rosenau, "Low-Cost Trigger-Pullers: The Politics of Policing in the Context of Contemporary 'State Building' and Counterinsurgency" (RAND, 2008), p 20.
6. Forrest E. Morgan, et al., "Dangerous Thresholds: Managing Escalation in the 21st Century" (RAND, 2008), p 43.
7. Ibid., p 172.

Chapter 13

1. David Marquand, "Playground Bully," from *Imperial Tense: Prospects and Problems of American Empire*, Andrew J. Bacevich, ed. (Chicago, IL: Ivan R. Dee, 2003), p 121.
2. Preamble and Article 1, Section 8.
3. Andrew J. Bacevich, "New Rome, New Jerusalem," from *Imperial Tense*, p 93.
4. Matloff, *American Military History*, p 102.
5. *The Collected Works of Abraham Lincoln*, edited by Roy P. Basler, Volume I, "Address Before the Young Men's Lyceum of Springfield, Illinois," (January 27, 1838), p 109.
6. Matloff, *American Military History*, p 161.

7.   Ibid., p 286.

8.   James Chace, "In Search of Absolute Security," *The Imperial Tense*, p 126.

9.   Matloff, *American Military History*, pp 323, 350.

10. Ibid., pp 409, 410, 414.

11. Ibid., p 540.

12. Ibid., pp 581, 616

13. "The Decline in America's Reputation: Why?" Subcommittee on International Organizations, Human Rights, and Oversight of the House Committee on Foreign Affairs (Washington, DC: U.S. Government Printing Office, June 11, 2008).

14. Ibid., p 1.

15. Ibid., p 4.

16. Ibid., p 18.

17. Ibid., p 20.

18. Ibid., p 21.

19. Stephen Melton, "Why Small Brigade Combat Teams Undermine Modularity," *Military Review* (July–August 2005), pp 58–63.

20. FM 3-24, Counterinsurgency (December 2006), p 1-2.

21. Clausewitz, *On War*, p 75.

22. Ibid., p 88.

23. Ibid., pp 87, 88.

24. Morgan, "Dangerous Thresholds," pp 155–57.

Chapter 14

Epigraphs. James Madison cited in Andrew J. Bacevich, *New American Militarism* (Cambridge, England: Oxford University Press, 2005), p 7; George Bush, Ibid., p 19.

1.   Adrian Goldsworthy, *Roman Warfare* (Washington, DC: Smithsonian Books, 2005).

2.   Ibid., p 34.

3.   Stephen Peter Rosen, "Imperial Choices," in *Imperial Tense*, pp 215, 216.

4.   Peter Bender, "The New Rome," in *Imperial Tense*, pp 85, 86.

5.   Ibid., pp 90, 91.

6.   William V. Harris, *War and Imperialism in Republican Rome, 327-70 BC* (Gloucestershire, England: Clarendon Press, 1992), pp 49–51.

7.   H. H. Scullard, *From the Gracchi to Nero: A History of Rome, 133 BC to AD 68*, 5th ed. (Routledge, 1992), p 49.

8.   Goldsworthy, *Roman Warfare*, pp 107, 108.

9.   Ibid., p 108.

10. Scullard, *From the Gracchi to Nero*, pp 69, 70.

11. Ibid., pp 77–83.

12. Plutarch, "Life of Julius Caesar"; Scullard, *From the Gracchi to Nero*, pp 134–37.
13. Ibid., pp 148–53.
14. Ibid., pp 210, 211.
15. Ibid., p 227.
16. Tacitus, *The Annals*, Book 1, chapter 7.
17. Interview with Gen. David McKiernan, *NBC Nightly News* (June 16, 2008).
18. Dwight David Eisenhower, Farewell Address to the Nation (January 17, 1961).
19. Chris Hedges, "What If Our Mercenaries Turn on Us?" *Philadelphia Enquirer* (June 3, 2007).
20. Ibid.
21. Ibid.
22. Memorandum to Members of the Committee on Oversight and Government Reform, Re Additional Information about Blackwater USA (October 1, 2007), pp 4, 5.
23. James Risen, "Army Overseer Tells Of Ouster Over KBR Stir," *New York Times* (June 17, 2008), p 1.
24. Dwight David Eisenhower, "The Chance for Peace," speech given to the American Society of Newspaper Editors, April 16, 1953.
25. Dennis S. Ippolito, "Budget Policy, Deficits, and Defense: A Fiscal Framework for Defense Planning (Carlisle Barracks, PA: Strategic Studies Institute, June 2005), p 3.
26. Ibid., p 9.
27. Budget figures from Travis Sharp, "U.S. Defense Spending, 2001–2009," www.armscontrolcenter.org, retrieved March 20, 2009.
28. Fiscal Year 2009 Budget of the United States, Historical Tables, Office of Management and Budget, pp 22, 54, 55, 133.
29. Ippolito, "Budget Policy," p 3.

Chapter 15

1. Michael R. Gordon, "Army Buried Study Faulting Iraq Planning," *New York Times* (February 11, 2008), p 1.
2. Guest speaker, U.S. Army Command and General Staff College, 2008. Unnamed due to CGSC non-attribution policy.

Appendix

1. Peter W. Chiarelli, "Winning the Peace: The Requirement for Full-Spectrum Operations," *Military Review* (July–August 2005), pp 4–17.

Index

Transvaal, South Africa, 88, 155
Treaty of Guadalupe Hidalgo, 39
Treaty of Paris, 46
Triple Canopy, 237
Triple Entente, 52
Truman, Harry, 3, 76, 203, 231
Tu, Colonel, 16
Tudjman, Franjo, 100

Umar, Mullah, 142
UN Security Council, 140
Uniform Code of Military Justice, 174, 237
Union Army, 42–45, 175
Union of South Africa, 88
United Nations (UN), 74, 75, 95, 98–100, 104, 112, 114, 140, 194, 206–208, 217
University of Maryland, 208
Urgent Fury, 84, 85
U.S. Army Center of Military History, 5
U.S. Army Civil Affairs and Psychological Operations Command (USACAPOC), 187
U.S. Census Bureau, 100
U.S. Central Command (CENTCOM), 6, 112, 113, 115, 117, 133, 175
U.S. Coast Guard, 210
U.S. Military History Symposium, 5
U.S. Southern Command, 85

Valley Forge, 37
Veracruz, Mexico, 39, 40
Vichy, France, 59
Viet Cong (VC), 16, 77–80
Viet Minh army, 76, 77
Vietnam, 2–4, 7, 8, 15–17, 42, 72, 76–81, 83, 90, 124, 128, 134, 136, 137 140, 148, 149, 153, 155, 170–172, 180, 182, 187, 190, 205, 213, 215, 231, 234, 235, 244
Vietnam War, 6, 7, 16, 24, 76–81, 127, 148, 165, 177, 203, 232
V-J Day, 64
Voice of America, 113
Vukovar, Croatia, 97

Wahhabist, 136, 184
Wall Street Journal, 105
War of 1812, 33, 38, 69, 215
War Powers Act of 1973, 232–234
Wardak, Abdul Rahim, 143
Warsaw Pact, 81, 82, 98
Washington, D.C., 3, 38, 41, 42, 59, 65, 66, 73, 76, 77, 79, 80, 107, 112, 113, 117, 121, 131, 134, 139, 159, 181, 182, 188, 189, 203, 206, 214, 235
Washington, George, 25, 36, 37, 174, 175, 202
Washington Naval Treaty, 203
Waxman, Henry, 237
Wehrmacht, 11, 83, 123, 175
Weigley, Russell, 148
Weimar Republic, 58
West Germany, 3, 58, 82, 187
Westmoreland, William, 2, 79, 80
Williams, Brian, 230
Winthrop, John, 201–203
Wolfowitz, Paul, 114
Wood, Leonard, 47
Wooster, David, 37
World Trade Center, 139, 230
World War I, 5, 6, 11, 14, 28, 51–55, 71, 148, 150, 175, 193, 196, 203
World War II, 1, 2, 4, 5, 14, 24, 27, 28, 31, 52–69, 71–76, 81–83, 85, 87, 88, 90, 91, 96, 117, 118, 120, 122, 123, 130, 138, 148–150, 152, 163, 165, 168, 170, 175, 177, 179, 180, 182, 184, 186, 188, 190, 193, 196, 203, 206, 211, 217, 229, 231, 234, 235, 240, 244
Wurmser, David, 133

"X Article," *Foreign Affairs*, 72
XXIV Corps, 65, 73

Yalta agreements, 71
Yalu River, 75
Yugoslavia, 93, 95–98, 104, 106, 124, 170; Civil War, 96, 97
Yugoslavian National Army (JNA), 97–99

Zepa, Bosnia-Herzegovina, 99